SUB-CONTRACTS UNDER THE JCT INTERMEDIATE FORM

PETER R. HIBBERD
MSc, ARICS, ACIArb

BSP PROFESSIONAL BOOKS

OXFORD LONDON EDINBURGH

BOSTON PALO ALTO MELBOURNE

First published 1987

British Library
Cataloguing in Publication Data
Hibberd, Peter R.
  Sub-contracts under the JCT
  Intermediate Form.
  1. Building — Contracts and
  specifications — Great Britain
  2. Sub contracting — Great Britain
  I. Title
  692'.8

ISBN 0–632–01869–0

BSP Professional Books
Editorial offices:
Osney Mead, Oxford OX2 0EL
  (*Orders*: Tel. 0865 240201)
8 John Street, London WC1N 2ES
23 Ainslie Place, Edinburgh EH3 6AJ
52 Beacon Street, Boston
  Massachusetts 02108, USA
667 Lytton Avenue, Palo Alto
  California 94301, USA
107 Barry Street, Carlton
  Victoria 3053, Australia

Set by V & M Graphics Ltd, Aylesbury,
  Bucks
Printed and bound in Great Britain by
  Mackays of Chatham, Kent

SUB-CONTRACTS
UNDER THE
JCT INTERMEDIATE FORM

ML

£22·50

Contents

contractor – Contractor's rights and obligations – Determination of employment of the sub-contractor

Preface

Relatively few books have devoted themselves entirely to building sub-contracts and even fewer have dealt with their provisions in any depth. This is perhaps surprising bearing in mind that for every main contract there is often a multitude of sub-contracts. Recognising this situation it was decided to fill part of that gap by covering the provisions of the IFC sub-contract NAM/SC in the belief that the JCT Intermediate Form of Building Contract will become a very commonly used form.

The approach to this book has been to provide a text that gives the reader an appreciation of the management and administration of sub-contracts in the context of the legal framework set out in NAM/SC. It is not a legal commentary but it does however discuss legal concepts and arguments where it is relevant to do so in order that the reader has a greater appreciation of the options, and the merit of those options, which are available in given situations. It is hoped that this book will be of use to everyone concerned with building sub-contracts under the Intermediate Form, regardless of whether their interest is because of a direct involvement or not. The reader will therefore find the subject matter to be an amalgam of the practical and the theoretical, bringing together the managerial and legal issues in a way that provides an understanding that will enable every day decisions to be made.

The clauses referred to are always those from NAM/SC unless stated otherwise for example clause 7.1 (IFC). Reference is usually made to the supervising officer and not the architect unless a specific purpose is intended in referring to the architect. No disrespect is intended to architects: this simply recognises the increasing use of non-architects in the administration of contracts.

Only passing mention is made of value added tax and the tax deduction schemes as it was felt that these are issues which deserve separate consideration and on which there are already a number of publications which are equally applicable to the NAM/SC provisions. Reference is made to all the other areas of the sub-contract provisions and it is hoped that this gives a reasonably comprehensive coverage of the subject and which also identifies a number of problems in the sub-contract so that they may in due course be remedied.

The text includes amendments made in November 1986 by the Joint Contracts Tribunal to the insurance provisions, but also includes discussions of the original provisions since these will apply to contracts let under the earlier version of the form.

My thanks are due to Miss Shirley Weeks for the extremely exacting task of preparing a manuscript from my own draft.

Peter R. Hibberd
Hanley Castle
December 1986

Chapter 1

The use of named sub-contractors

Definition of naming

The use of sub-contractors in building works is a common enough occurrence and has been for many years. Yet in spite of this, difficulties regularly arise with regard both to their selection and to their performance under the contract. Sub-contracting is fraught with problems because frequently one has no clear idea of what it is that one is seeking to achieve. This can be seen by reference to the range of terms used to describe sub-contractors, such as:

- named sub-contractors
- nominated sub-contractors
- domestic sub-contractors
- domestic sub-contractors as prescribed by clause 19.3 of JCT 80 (that is, by reference to a select list)
- specified sub-contractors
- approved sub-contractors
- specialist sub-contractors, and just plain simple sub-contractors

This range of terms to describe sub-contractors has largely come about as a consequence of the use of sub-contractors in a variety of contractual arrangements. The use of a qualifying term such as 'domestic', 'named' or 'nominated' is an attempt to indicate not only the nature of the contractual arrangement in which the sub-contractor finds himself but also the extent of the design team's control and employer's involvement. Unfortunately, the terms used do neither particularly well and it is therefore worthwhile attempting an explanation of the term 'named sub-contractor'.

It is clear that in the general sense any sub-contractor who is selected by the design team could be said to be named. This being so, one can see that a nominated sub-contractor, and even a domestic sub-contractor where the contractor has been restricted in his choice, could be described as a named sub-contractor. However, when the term was used in the JCT Intermediate Form, IFC 84, it was intended to describe a very specific type of sub-contract relationship, being neither nomination nor domestic.

IFC 84 does not define a named sub-contractor and only provides in clause 3.3.1 as follows:

> 'where it is stated in the Specification/Schedules of Work/Contract Bills that work described therein for pricing by the Contractor is to be executed by a named person who is to be employed by the Contractor as a Sub-Contractor...'

and also in clause 3.3.2(a) where it gives the right for an instruction against a provisional sum to require work to be executed by a named sub-contractor.

This is in contrast to JCT 80 where a comprehensive definition of nomination is contained within clause 35.1:

> 'Where...the Architect/Supervising Officer has, whether by the use of a prime cost sum or by *naming a sub-contractor*, reserved to himself the final selection and approval of the sub-contractor to the Contractor who shall supply and fix any materials or goods or execute work, the sub-contractor so *named* or to be selected and approved shall be nominated ...'

Upon analysis it is therefore apparent that a named sub-contractor as used under IFC 84 would be considered nominated if used under JCT 80. The use of terms in this way is not helpful and is in fact misleading because one sees that the use of the term 'named sub-contractor' has a different meaning under these two contracts.

A named sub-contractor under IFC 84 is a sub-contractor who is stated as such either in the contract documents or by way of an instruction. Clause 3.3 provides the contractual framework which governs the use of named sub-contractors. The work of the named sub-contractor is to be priced by the main contractor as compared with the requirement to include either a prime cost sum or other sum stated by the sub-contractor.

The contractual relationship of the named sub-contractor with the main contractor is purely domestic in that neither the employer (that is, the client) nor the supervising officer is concerned with the contract between them. They are not contractually concerned with matters such as how much the sub-contractor should be paid, has he been paid, direct payment where the main contractor has defaulted, or the granting of an extension of time to the sub-contractor. Therefore, the sub-contract conditions NAM/SC more closely resemble the DOM/1 contract, which is used for domestic sub-contractors under JCT 80, than NSC/4 or 4a which are used for nominated sub-contractors.

A distinction between a domestic sub-contractor and a nominated sub-contractor can therefore be drawn at the point where the supervising officer and/or employer are no longer involved in the operation of the terms of the contract between the main contractor and sub-contractor, and no longer involved with other matters directly affecting the position of the sub-contractor once in contract with the main contractor.

However, the distinction between a domestic sub-contractor and a named sub-contractor is not too obvious. It is convenient and straightforward to simply say that a named sub-contractor is also a domestic sub-contractor unless he can be described as nominated. But it cannot be said that every domestic sub-contractor is also a named sub-contractor.

Although a named sub-contractor is a domestic sub-contractor, treating them as one and the same fails to recognise that a distinction is possible and may be desirable. A main contractor is generally recognised as having full responsibility in all respects for a domestic sub-contractor. This includes responsibility for a failure of the sub-contractor which delays completion of the main contract work. However, under IFC 84 a main contractor is relieved of some of these responsibilities when employing a named sub-contractor.

Therefore, one may say that a sub-contractor is a named sub-contractor when the designer specifies the firm, when the operation of the sub-contract is in no way affected by the supervising officer or employer but where the main contract regulates some of contractual obligations which would otherwise fall upon the main contractor. This compares with a nominated sub-contractor because the sub-contract here would contain provisions which enable the supervising officer or employer to affect the operation of the contract.

For example, clause 12.2 of NSC/4 states:

'Provided that the Contractor shall not be entitled so to claim unless the Architect ... shall have issued to the Contractor ... a certificate ...'

Furthermore, it can be contrasted with a domestic sub-contractor because here the designer would not be specifying the form of contract to be used.

The foregoing definitions are sensible in that they are workable and give us a clear idea of what we are seeking to achieve. It is therefore unfortunate that JCT 80 refers to named sub-contractors as nominated and that clause 19.3 of JCT 80 introduces a naming aspect (by specifying the list of sub-contractors) into what are clearly described as domestic sub-contractors.

Domestic sub-contractors

Most forms of building contract make reference to general sub-contracting, no doubt in recognition of the fact that sub-contractors and sub-letting of the work by the main contractor are generally accepted practices. For example, clause 3.2 (IFC 84) provides:

> 'The Contractor shall not sub-contract any part of the Works other than in accordance with clause 2.3 without written consent of the architect/supervising officer whose consent shall not unreasonably be withheld.'

It is perhaps as well that most standard forms of building contract make such provision because the right to sub-contract in the absence of such a provision and without the approval of the employer is dubious. It is dubious because the precise legal position is debated, and this can be illustrated as follows. To quote Neil Jones and David Bergman in their book. *A Commentary on the JCT Intermediate Form of Building Contract*:

> 'Sub-contracting of contractual obligations, i.e. vicarious performance of the actual work, is permitted under the general law unless the obligations are personal in nature.'

Compare this with the view expressed at length by John Parris in his book *Default by Sub-Contractors and Suppliers* where he argues that contractors have no general right to employ sub-contractors.

The latter is to be favoured as a correct interpretation of the law, and one should not be misled by the apparent support given by I. Duncan Wallace in *Building and Engineering Contracts* to the view that the law permits the vicarious performance of contractual liabilities, except in the case of personal contracts, since he later states that vicarious performance will not be permitted if the result will be to alter or prejudice the obligations or rights of the other party. It is upon this issue that the authors previously referred to disagree, but it seems very clear that in law the main contractor would have to establish a right to sub-contract. It would not be for the building employer to prove that no right exists in circumstances where the main contractor has sub-contracted work without approval.

From this discussion one can see that provisions which allow sub-contracting are recognising this difficulty, but again it is easy to misconstrue the legal effect of such a clause as 3.2 (IFC 84). Superficially, it seems to be establishing the position which is expressed by John Parris as the position in law – that is, you cannot as of right sub-contract without approval of the employer. However, this clause does not provide for approval by the employer; it simply requires the consent of the supervising officer in respect of the contractor's proposal for sub-contracting. This consent is given by the supervising officer in his capacity as agent for the employer and such consent must be given unless the supervising officer can establish good grounds for refusal. No doubt a very good ground for withholding consent would be that the work requires a particular skill for which the contractor was himself chosen. One can therefore see that clause 3.2 effectively modifies the common law position in that the clause gives the contractor a right to sub-contract unless the supervising officer has grounds to prevent it.

Sub-contractors who are selected in this way by the main contractor are known as domestic sub-contractors. However, domestic sub-contractors may also come about by a very different procedure which is adopted by JCT 80. Clause 19.3 (JCT 80) provides:

'Where the Contract Bills provide that certain work measured or otherwise described in those Bills and priced by the Contractor must be carried out by persons *named* in a list ...'

The provision of a list of sub-contractors under this procedure means in effect that the sub-contractors are named, but such naming is not referring to the final selection, which is reserved for the contractor. Therefore, because the contractor makes the final choice, albeit from a restrictive list of firms, it is generally considered that this is not naming as such.

Specified, approved and specialist sub-contractors

In order that a complete picture of the terms used to describe sub-contractors is given, it is worth briefly referring to specified, approved and specialist sub-contractors.

A *specified sub-contractor* is generally considered to be one referred to in the contract documents. Clearly, the sub-contractor is named but he may be either domestic or nominated in nature, depending upon the contractual arrangement.

An *approved sub-contractor* is a sub-contractor who has been accepted by the supervising officer, following a request by the main contractor to perform work, which in the contract documents has not been reserved or specified to be carried out by a sub-contractor. In other words, it is the approval of a proposed sub-letting of the work.

However, even this fairly straightforward definition is threatened by the proposals for the third edition of GC/Wks/1 (currently in draft form only) in that an approved sub-contractor is defined as:

'any one of two or more persons named in a list provided or to be provided ... as being approved for the execution or provision of any particular work ...'

Nevertheless, both definitions do intend the sub-contract to be domestic as does the provision contained in clause 19.3 of JCT 80. The GC/Wks/1 proposals resemble the JCT 80 provision but the latter contract does not expressly refer to them as approved.

Specialist sub-contractor is a term often reserved to describe a sub-contractor who provides a specialist service such as heating, ventilating or electrical, as compared with the sub-contractor who executes general building work. The term generally has no contractual significance but once again the proposals for the third edition of GC/Wks/1 will, if adopted, affect the position, A specialist sub-contractor is here defined as:

'a single person named in the contract or in any supervising officer's instruction as the person who is as a sub-contractor ... to execute or provide any particular work ...'

The intention here is to adopt a similar approach to naming as prescribed by IFC 84, and the specialist sub-contractor can be considered as named for the intention is clearly to move away from the traditional approach of nomination.

Conditions required to be included in domestic sub-contracts

The main contractor under IFC 84 is required to provide in any sub-contract with a domestic sub-contractor arising by virtue of the operation of clause 3.2 clauses which have the following effect:

(1) the employment of the sub-contractor to be determined immediately upon the determination of the main contractor's employment.
(2) unfixed materials and goods, which have been delivered to or placed on or adjacent to the works by the sub-contractor shall not be removed without the consent of the contractor (the contractor himself having to receive consent from the supervising officer).
(3) where the value of such materials or goods has been paid for by the employer, those materials or goods become the property of the employer and the sub-contractor is estopped from claiming otherwise.
(4) where the value of such materials or goods have been paid for by the contractor before their value has been paid by the employer to the contractor, they become the property of the contractor.
(5) the operation of the clauses required is without prejudice to any property in off-site materials or goods passing to the employer.

At present there is no standard form of contract for domestic sub-contractors under IFC 84 and therefore the contractor should take care to ensure that these clauses are included in the sub-contract. Failure to include these clauses in the sub-contract is a breach of contract which it has been suggested may enable the employer under clause 7.1(d) to determine the contractor's employment. Upon reading clause 7.1(d) one is left with the view that it must apply to the failure to include appropriate clauses in the sub-contract.

This view is held by Neil Jones and David Bergman in *A Commentary on the JCT Intermediate Form of Building Contract* but it is difficult to be adamant about this point because although clause 7.1(d) refers to clause 3.2 it does not refer specifically to clauses 3.2.1 or 3.2.2. One might normally accept that as these are sub-clauses of clause 3.2 they must therefore be deemed included particularly as 3.2 makes reference to the requirements. But having considered the circumstances in which clause 7.1(d) could operate one is left in doubt. To enable a determination to take place the contractor must continue such default for 14 days after receipt of a notice specifying the default. It is clear that once the default has occurred the contractor is going to be unable to resolve the problem because the sub-contractor is most unlikely to agree to the inclusion of further terms which are to his disadvantage.

Furthermore, the default can only occur once the sub-contract has been entered into; it cannot occur prior to the contract being formed. Therefore one can see that the implementation of the procedure under clause 7.1(d) does not give the contractor any possibility of redressing his breach and one is left in doubt as to whether it is intended to apply to clauses 3.2.1 and 3.2.2. Bearing in mind this and the fact that no such notice of determination shall be given unreasonably or vexatiously one is tempted to suggest that it would be unwise for an employer to rely upon this clause. It would seem preferable to sue for damages for the breach of clause 3.2.

A simple amendment to the drafting of clauses 3.2, 3.2.1 and 3.2.2 and 7.1(d) would be of assistance in order to precisely establish whether it is intended to provide that a main contractor may have his employment determined because of the failure to incorporate such terms. Otherwise, it remains a possibility that the courts would interpret the words 'if the contractor shall continue such default for 14 days' in clause 7.1(d) as being limited to only those situations where the contractor has the ability to put right his default. If this argument prevailed one would then only be able to implement the determination of the main contractor's employment where there was a repeat of such default – that is, having failed to include these terms in one sub-contract, then subsequently repeating the default with another sub-contract when notice of the former default had already been received from the supervising officer.

Reasons for using sub-contractors

There are basically two aspects from which this can be viewed, namely that of the main contractor and that of the building employer. From a main contractor's point of view sub-contracting is convenient because:

(1)   it reduces the risks of the contractor (this is particularly so when the sub-contractor is nominated).

(2) it is generally cost effective to employ domestic sub-contractors rather than to employ their own direct labour.

Against this the contractor may have to suffer the effect of the design team's ability, through the nomination process, to delay design decisions and therefore cause problems of co-ordination, or to suffer the restrictions where a sub-contractor is named. Naming restricts the ability of the main contractor to deal commercially once the main contract has been entered into.

On balance, however, the contractor is generally well pleased to be able to use sub-contractors whether they be domestic, named or nominated and therefore one is unlikely to see much pressure from this quarter to move away from the use of sub-contractors. In fact the reverse is probably true, for we have seen a move away from main contractors maintaining large numbers of directly employed persons and a move towards the use of sub-contractors. However, there is some evidence that contractors regret the move away from nomination to other forms of sub-contracting, no doubt on account of having to accept a higher level of risk.

Not surprisingly a different picture emerges when looking at sub-contracting from the building employer's – that is, the client's – viewpoint. Traditionally, the design team have presented the client's view but it is becoming clear that this may not be totally appropriate.

Firstly, it may be argued that if a client enters in to a contract with a building contractor he ought to be assured that the contractor he has selected will perform the whole of the works. Today this would seldom be the case for many standard forms of contract allow for the supervising officer to permit the contractor to sub-let a portion of the works. It seems that generally speaking sub-contracting is an acceptable practice and it is argued that so long as the client achieves that for which he has contracted there is little point of holding to the idea that the main contractor should perform the whole works, notwithstanding the fact that the main contractor remains responsible for the work performed by sub-contractors. Following this logic through it could be argued that if the end result is all that matters and if the main contractor remains liable then surely there can be no objection to the principle of sub-letting. Nearly all building contracts, although providing for sub-contracting, usually ensure that the supervising officer has the ability to prevent sub-letting where it is reasonable to do so. The provisions therefore appear to accept the principle of sub-letting but not necessarily sub-letting to the firm proposed by the contractor to carry out the work.

The circumstances in which sub-letting may be refused are various but may include the following:

- it is for labour only
- a poor standard of workmanship on earlier projects
- a poor record in respect of time performance
- doubt concerning financial stability

The building employer is generally well protected from a legal point of view in respect of these issues but is also requiring protection from the management problems which may ensue. In fact one could argue that it is the supervising officer who is really protecting himself from a potential workload for which there will be no remuneration.

Secondly, and no doubt more importantly, a majority of designers have argued that the ability to choose (as compared with to approve) the sub-contractors with which the main contractor must contract is essential in some circumstances. This, they argue, is necessary to make certain that adequate expertise is available to ensure the achievement of appropriate standards. This is particularly relevant when the nature of the work requires a specialised design involvement. Few people would argue with this but they would argue about how it is to be achieved. It would not be appropriate in a book of this nature to investigate this problem in depth but it is worth discussing the move away from nomination.

Nomination served the purpose of the above objective but it possessed two basic flaws in the way it was operated: firstly, it left the building employer with a high degree of risk in respect of that work, and secondly, the general approach of using prime cost sums to cover nominated sub-contractors meant that many projects were inadequately designed or ill conceived at pre-contract stage, the consequences of the latter falling upon both the building client and the building contractor but seldom upon the design team.

Therefore, it became apparent to building employers and the more enlightened members of the design team that if this ability to choose one's sub-contractors was to be maintained, the way it was operated had to be changed. This resulted in two separate approaches: by expanding the choice as now provided for in clause 19.3 of JCT 80, or by naming the sub-contractor in the contract documents and at the same time moving more responsibility onto the contractor. The former does not entirely maintain final choice and therefore where this is deemed essential a named sub-contractor is the answer.

The named sub-contractor being named in the contract documents ensures that the design team do their work at the appropriate stage. Furthermore, because the main contractor knows at tender stage with whom he is going to contract it is not thought unfair to impose upon him greater responsibility, thus relieving the building employer of some of his burden.

Unfortunately, the further ability to name a sub-contractor in a supervising officer's instruction disrupts the apparent equilibrium and enables the designer to put off making a decision. It is accepted that there are occasions when this is inevitable. As a consequence it is believed that where such an instruction introduces a sub-contractor the responsibilities of the main contractor should perhaps be modified, but not to the same extent they would be if the sub-contractor were to be nominated. One can therefore see that building employers are now having an impact upon how buildings are procured but it is also apparent that the design team is still endeavouring to ensure that it has the greatest flexibility in which to work, if possible at the expense of the others.

Chapter 2

The naming provisions of the Intermediate Form

Generally

There is no provision in IFC 84 for nominated sub-contractors or suppliers as defined in JCT 80 but there is provision for named sub-contractors. Before considering the naming provisions contained in IFC 84 it is beneficial to view briefly the scope of this form of contract. This will provide an understanding of the circumstances in which the naming provisions were anticipated to be used, and when considering the contract terms they can be seen and appraised against this background.

The Intermediate Form is designed to be used where the proposed building works are:

(1) of simple content involving basic trades and skills.
(2) without complex services or other complex specialist work.
(3) adequately specified or specified and billed.

A statement regarding the use of the form is endorsed on the back of the contract itself. However, further guidance as to the form's use is contained in Practice Note 20 (revised 1984) which provides that the form would normally be used on contracts where:

(1) the contract period is not to exceed 12 months' duration.
(2) the contract sum is not to exceed £250,000 at 1984 prices.

It should be noted that the foregoing are guidelines only and not mandatory requirements. Therefore, one is free to make a judgment in respect of each project and it has already been reported that the form is being used on contracts valued at £6 million. The reason for using the form in such circumstances is presumably that it meets the first three requirements as stated above and, as it does, the duration and cost are of little significance. One cannot be sure that this is sound because weak contract clauses are generally only ever tested once the stakes make it a worthwhile proposition. Nevertheless, it is clear that the form is being used on very large contracts and no doubt will continue to be used on occasion in preference to JCT 80 even on contracts where the specialist involvement is beyond that anticipated when IFC 84 was prepared.

The Intermediate Form is also designed to be used with or without bills of quantities and therefore may be said to be both a with and a without quantities edition. In either case drawings are required, but it is evident that the way the naming will take place will be variable and dependent upon the relevant contract document. Distinct alternatives, which in themselves contain options, are provided for and referred to in the second recital. The alternatives are:

(1) the use of drawings together with a priced specification, schedule of work or bills of quantities.
(2) the use of drawings together with a specification.

The priced specification, schedule of works or bills of quantities is known as the 'priced document'. Where these are not used the priced document will be either a contract sum analysis or schedule of rates. This need not determine the type of sub-contract documentation but it probably should, and this is discussed further later in this chapter.

The time to name

The design team are confronted with a number of decisions in this respect. They must decide whether they wish to name a sub-contractor or whether they are to leave it to the main contractor. It is not appropriate in this book to rehearse the arguments for and against the designer reserving for himself the choice of who performs the work. However, it is appropriate to say that naming should only be used where it can be justified against sound criteria (see page 8), not simply because it is expedient.

Furthermore, it must be appreciated that the provisions of clauses 3.3.1 to 3.3.7 regarding naming do not apply to local authorities or statutory undertakers when executing work solely in pursuance of their statutory rights or obligations. If one is going to name one must decide at what stage and in what manner this is to be done.

In IFC 84 named sub-contractors may be brought about in two very distinct ways: by naming in the contract documents, or by naming in an instruction given for the expenditure of a provisional sum. Therefore, one may either name pre-contract or post contract but one's choice is usually determined by the time available. In order to provide expressly for a named sub-contractor in the tender documents it is necessary to ensure that the lead in time is available to do this and, if this is the case, that it is used appropriately. In practice, this presents similar problems to sorting out all one's nominations before going out to tender for the main contract works. Therefore, it is most likely that many sub-contractors who might or possibly should have been named in the tender documents will simply be covered by the use of a provisional sum. The use of a provisional sum should be avoided for the benefits of pre-contract naming over post-contract naming are manifest and are discussed in this chapter (see pages 24–25).

Pre-contract naming

Selection of tender

In order for a sub-contractor to be named, whether pre-contract or post contract, it is necessary to secure a tender on the Form of Tender and Agreement NAM/T. The detail of this form, its operation and effect is dealt with in detail in chapter 3.

Once it has been decided to name a sub-contractor in the tender documents it is necessary to secure an acceptable tender from a specialist. The intention is that only one firm is named for any specific part of the works and therefore it is necessary to select one tender only. Whether the design team choose to go out to a range of specialists or negotiate with one or more generally depends upon the extent of competition required and available and the level of design involvement.

The employer is required to give to the contractor a full description of the work for which a named sub-contractor has been chosen. This is clearly stated in the first recital of IFC 84:

'and in respect of any work described and set out therein for pricing by the contractor and for execution of which the contractor is required to employ a named person as sub-contractor in accordance with clause 3.3.1 of the Conditions annexed thereto, has provided all of the particulars of the tender of the named person for that work in a Form of Tender and Agreement NAM/T...'

The inclusion of this information in the tender documents will enable the contractor to assess the work and to submit *his* price for the sub-contract work to the employer. The contractor's price may – and will, in all probability – be different to that quoted by the sub-contractor because it takes account of other costs. This feature clearly distinguishes naming from nomination and the benefits of a competitive sub-contract price where the sub-contractor is named need not necessarily be passed on to the building employer.

Incorporation of naming in the tender documents

The way a named sub-contractor is provided for in the tender documents is largely determined by the choice of the contract documents to be used. If a specification is to be used, all that is required is an item giving a full description of the works and a reference to NAM/T. NAM/T submitted by the sub-contractor who is now named should be appended to the specification. This information enables the contractor either to price the specification or alternatively, where no priced specification is required, to use it as a basis for establishing the contract sum analysis.

It may be appropriate to give a number of items in the specification to cover the work to be executed by the specialist. However, one should ensure that the items that the main contractor is required to price are compatible with the sub-contract documentation upon which the named sub-contractor has tendered and, hopefully, priced. To provide otherwise would put the main contractor to additional work that could be avoided.

However, it is worth mentioning that there is nothing to prevent a specification for the main contract works containing within it a list of items with quantities in respect of named sub-contract work. This is made possible and provided for in clause 1.2:

'where and to the extent that quantities are contained in the specification ... and there are no contract bills, the quality and quantity of the work included in the contract sum for the relevant items shall be deemed to be that which is set out in the specification ...'

Nevertheless, although not impossible it is difficult to conceive a situation where this would be appropriate, particularly bearing in mind the type of project for which the form is considered suitable.

Where bills of quantities are adopted the matter of including named sub-contractors is generally somewhat more complex. The work to be covered may be:

(1)   measured out in detail.
(2)   included as an item or items to embrace the whole work in a similar way to that generally provided in a specification.

If bills of quantities are adopted those bills are required to be prepared in accordance with the Standard Method of Measurement of Building Works 6th Edition, unless otherwise expressly stated. Therefore, if one wishes to adopt the second approach mentioned above it is necessary to expressly state that the item(s) are not measured in accordance with SMM6.

Whichever approach is adopted it is again abundantly clear that the main contract documentation should be compatible with the sub-contract documentation. If this is not achieved it can cause a good deal of additional and/or unnecessary work for both the main contractor and the sub-contractor. Therefore, those responsible for producing the contract documentation should ensure that this is achieved and sub-contractors when pricing should price accordingly, bearing in mind that the main contractor needs to use his prices directly. Once one appreciates the need for compatibility of documentation, particularly where bills of quantities are adopted for the main contract, it is easy to see the difficulty with which one is confronted when this is not achieved.

The time factor in producing measured items for sub-contract tenders and securing tenders before the main contract bill can be compiled presents a very real problem. This problem is exacerbated when the sub-contractor is required to design the sub-contract works because until the work is designed it cannot be measured. It will therefore not be too surprising to see the second of these methods, i.e. including only one or a limited number of items, becoming the norm.

The RICS Quantity Surveying Practice and Management Committee has prepared a guidance note as follows:

'Works to be executed by named persons who will become sub-contractors should be included in bills of quantities on the basis considered most appropriate, except that prime cost sums are inadmissable.'

In considering what is the most appropriate way of including them one is left with the belief that it will be more a function of time than the genuine merits of whether or not a price breakdown in bill form is necessary for the purposes of valuation.

If the work is measured out in detail the contractor is required to price each and every item. When inserting his price for the work he should consider and include as necessary for his profit, special attendance, general attendance and any costs which he considers necessary. It is assumed that the contractor will use the named sub-contractor's quoted prices as the basis for his own but there is nothing that requires him to do so and, as previously mentioned, there is no guarantee that the building employer will secure the benefits of the named sub-contractor's price. The client can of course expect that the named sub-contractor will execute the work although, as will be seen, there is no guarantee in this respect either.

Although it is anticipated that the contractor would usually quote an all inclusive rate for each item, there is nothing to prevent him quoting net items and pricing items which have been separately billed in respect of such items as profit, special attendance, general attendance and other costs. This is a similar approach to that of nomination and, although not generally intended, it is possible so long as one recognises that such items should be appropriately qualified by expressly stating that these items are not measured in accordance with SMM6.

Where the named sub-contract work is included as one or a limited number of items in the bills of quantities, it has the advantage of simplicity. Unfortunately, however, it also may have severe disadvantages in that it does not provide a suitable framework for valuing either the works in progress or the cost of variations.

Entering into a sub-contract

The purpose of pre-contract naming is to achieve the selection of a specialist whilst passing much of the risk for carrying out that work on to the main contractor. As the contractor is aware at the time of his tender which sub-contractors are to be employed, no right of reasonable objection is provided for in the main contract. The selected contractor is required under clause 3.3.1 (IFC 84) to enter into a sub-contract with the named sub-contractor not later than 21 days after entering into the main contract.

If the contractor fails to enter into a sub-contract as provided for in the contract documents, he is in breach of contract except where he can avoid this obligation on the grounds that he is unable to enter into a sub-contract in accordance with the particulars given.

Being unable to enter into a sub-contract should not be construed widely for the contractor's inability to conclude a sub-contract must be on account of the particulars specified. This in effect means the sub-contractor is either no longer prepared to enter into a contract in the terms of his original offer, or alternatively is no longer able on account of liquidation or bankruptcy. Additionally the contractor may not be able to accept the terms of the sub-contractor's offer because they are already out of date owing to a delay that has occurred between obtaining the sub-contract and letting the main contract. Where this occurs the contractor must immediately inform the supervising officer and specify which of the particulars has prevented the execution of a sub-contract.

It would seem that the contractor is not required to notify the supervising officer as soon as a problem occurs, notwithstanding the use of the word 'immediately', but only to do so once he has been prevented from entering into a sub-contract, and this would not normally be considered finally until the 21 days have elapsed. Where the contractor knows he is unable to enter a sub-contract because the terms are already out of date he should immediately notify the supervising officer.

Although the contractor is under an obligation to conclude a sub-contract, the sub-contractor is not. The reason for this is that the main contractor is bound by the terms of the contract he has entered into whereas the sub-contractor has only made an offer which in law can be withdrawn at any time before it is accepted – except, that is, where there is an agreement to keep the offer open for a given period. Such an agreement would require consideration unless it was made under seal. No such agreement is made when a sub-contractor tenders under NAM/T.

It would seem that generally sub-contractors would not wish to create problems in this respect and would feel that the commercial pressures are sufficient in themselves to keep the offer open. Nevertheless, they are able to withdraw at will and may also take advantage of the situation to increase their tender. This may happen because there is a genuine change in the circumstances which prevailed when they made their offer. Such a change could occur because the building employer has delayed the main contract, because of economic conditions i.e. the bank base rate has altered, or for a variety of other reasons. Delay caused by the building employer may be a justifiable reason but the other reasons, which are no more than commercial risks, are somewhat debatable. Nevertheless, the sub-contractor is perfectly entitled to withdraw his offer regardless of reason.

A supervising officer confronted with the situation where the sub-contractor is seeking to amend his offer and by so doing is preventing the execution of a sub-contract has three possible courses of action under the main contract:

(1) he may change the particulars so as to remove the impediment; this in practice will often mean agreeing to an increased price.
(2) he may omit the work.
(3) he may omit the work from the contract and substitute a provisional sum.

Although the supervising officer has a choice he will in reality often be restricted to agreeing to the increased price sought on account of the delay which will otherwise ensue. The named sub-contractor is in a strong position and it is only the fact that such demands may sour future work that will redress the balance.

Changing the named sub-contractor

Even though a sub-contractor has been named in the main contract documents he cannot be certain that he will secure the sub-contract work. The main contract may never be let, but even where it is the supervising officer has the ability under clause 3.3.1 (IFC 84) to issue an instruction to the effect that the named sub-contract work is to be carried out by a person other than a person named in the contract documents. This instruction may be issued at any time prior to the date when a sub-contract is entered into between the main contractor and original named person, and a named sub-contractor should take notice of such a facility.

Although this is clearly the intention of clause 3.3.1 the clause does not actually say this because it refers to the 'date so notified' by the contractor, thus implying that until he does so the supervising officer has a right to change his mind.

The supervising officer may attempt to change his mind between a sub-contract being entered into and the date of that sub-contract being notified by the contractor, in accordance with clause 3.3.1. However, if a sub-contractor is confronted with the position whereby the main contractor is instructed to enter into *another* sub-contract for the same work, the sub-contractor would have a remedy in law against the contractor and possibly the supervising officer if such an instruction led to the repudiation of the contract between the first named sub-contractor and the main contractor.

This, of course, should not happen because if a main contractor receives such an instruction and a sub-contract has been created then the main contractor would be advised not to act upon the instruction and to ask the supervising officer to withdraw it. A supervising officer who subsequently discovers that he would prefer to name another sub-contractor may not wish to withdraw his instruction but he would be unwise to act otherwise.

It seems self-evident that no instruction of the supervising officer should be able to remove the whole of the sub-contract works from a sub-contractor once a sub-contract has been entered into. Unfortunately the words used in the main contract are not so clear. The words as they stand are not good enough and should be amended.

If the supervising officer issues an instruction to the effect that work is to be carried out by a person other than a person named in the contract documents it seems there is nothing, save an existing valid sub-contract, that restricts him to naming another sub-contractor. The express provision contained in clause 3.3.1 (IFC 84) may also enable the supervising officer to require the contractor to execute the work himself because such an instruction has the effect of omitting the work in the contract and substituting a provisional sum. There is no express requirement that the provisional sum must be expended by naming another sub-contractor.

Admittedly, an instruction under clause 3.3.1(c) is required to be dealt with in accordance with clause 3.3.2 but this clause is not as precise as it should be. It reads:

> 'In an instruction as to the expenditure of a provisional sum under clause 3.8 the supervising officer *may* require work to be executed by a named person ...'

It therefore appears that a supervising officer may have a right to require the main contractor to execute named sub-contract work which is provided for in the contract. If this is so then it contrasts with the position of the main contractor under SFBC JCT 80 where no such right exists. The position under JCT 80 is stated in clause 19.5.2:

> '... the contractor is not himself required to supply and fix materials or goods or to execute work *which is to be carried out by a nominated sub-contractor*'

Unfortunately, even these words are not as precise as they could be because they might preferably say 'which is provided in the contract documents to be carried out by a nominated sub-contractor'. Nevertheless, the position in law is quite clear, particularly in the light of the case of *Bickerton* v. *North West Metropolitan Hospital Board* (1970) where it was established that the architect must, under JCT 63, renominate and, furthermore, that the main contractor would be in breach himself if he without approval executed work set aside for a nominated sub-contractor. This does not mean that a main contractor can never be required to execute work set aside for a sub-contractor because the contract conditions may provide for just that possibility. It is submitted that in order to achieve this position the contract provisions would need to be clear and unambiguous.

A main contractor under the Intermediate Form may be required to execute work set aside for a named sub-contractor. There is provision that named sub-contract work may be omitted and for a provisional sum to be substituted, with no conditions as to how the provisional sum should be expended. Clause 3.8 simply requires the supervising officer to issue instructions as to the expenditure of any provisional sums. It could be further argued that instructions as to the expenditure of a provisional sum will normally require the main contractor to execute the work, with clause 3.3.2(a) (IFC 84) only providing for the exception whereby a named sub-contractor is required. To put it another way, clause 3.3.2(a) allows for a named sub-contractor to be used under an instruction as to the expenditure of a provisional sum. In spite of this weight of argument it is still probably intended that where a named sub-contractor is to be replaced it is implied that he will be replaced by a named sub-contractor by way of an instruction as to the expenditure of the provisional sum.

Because clear words are not used it would seem prudent to ensure that once a named sub-contractor is identified in the contract document he is then used. Where this is not possible because circumstances change, another named sub-contractor should be used in preference to the main contractor. If, however, it is decided to use the main contractor for such work his approval should be obtained before instructing him to execute it. This would then avoid any possible argument on this point. However, if the contractor refuses to co-operate one is still left with a problem, and although one favours the view that the supervising officer has the right to require the contractor to execute such work it would in these circumstances be preferable all round to name another sub-contractor.

Post-contract naming

Clause 3.3.2(a) (IFC 84) enables the supervising officer to name a sub-contractor during the post contract stages for it provides:

'... instruction as to the expenditure of a provisional sum under clause 3.8 the supervising officer may require work to be executed by a named person ...'

In order to secure post contract naming a provisional sum is required. Therefore, where one anticipates using a named sub-contractor but one is not in a position at tender stage to name that person a provisional sum is included in the tender documents. As already discussed, it is also possible for the provisional sum itself to be created at post contract stage but this does not materially affect the way that a sub-contractor is named at this stage.

The only justification for providing in the contract documents for post contract naming is that time is not available to adequately provide all the information for pre-contract naming. In practice this is often the position, so although pre-contract naming is advantageous it is not always possible. The problem of time cannot always be avoided and procurement methods must accommodate these matters as best they can rather than imposing upon the client a framework which does not enable his objective to be achieved. Therefore, it is clear that post contract naming will be a fairly frequent occurrence and no doubt in many cases rightly so.

However, one should not post contract name where it can be avoided and the supervising officer should not use this facility simply because it is expedient from his point of view to do so. Such a use is an abuse of the facility and supervising officers should not adopt post contract naming on account of their own management deficiencies.

Selection of tender

In order for a sub-contractor to be named at post contract stage it is still necessary to secure a tender on NAM/T. The supervising officer in his instruction to the contractor under clause 3.3.2(b):

'shall incorporate a description of the work and all particulars of the tender of the named person for that work in a Form of Tender and

Agreement NAM/T with sections I and II completed together with the Number Documents referred to therein.'

Therefore, it is necessary for the supervising officer to go out to tender and to select a tender from those submitted. The documentation used to obtain the sub-contract tenders will be dependent upon the scope and nature of the work. As only a provisional sum is used in the main contract documents there is no need to ensure that there is compatibility with such documentation, as is the case when using pre-contract naming. In pre-contract naming the contractor is required to price the sub-contract work using the sub-contract price as a basis, whereas in post contract naming the value of the named sub-contract work will be valued in accordance with clause 3.7 (IFC 84).

It is possible that the contractor may be asked to submit his price for agreement with the employer in respect of the sub-contract work, as is provided for as an option in this clause, but this is not an essential requirement.

Entering into a sub-contract

In post contract naming the contractor is unaware at the time when he enters into the main contract who the named sub-contractor will be or under what conditions it is proposed to employ him. Therefore, unlike the pre-contract naming situation the main contractor is given the opportunity to make a reasonable objection. An objection may be made on a variety of grounds, such as the programming of the works or past experience of the sub-contractor with regard to quality and/or speed in operation. Unfortunately, what constitutes a reasonable objection and in what circumstances it will be sustainable is uncertain.

The main contractor is required under clause 3.3.2(c) to enter into a sub-contract with the person named in an instruction unless he has made a reasonable objection within 14 days. If the main contractor fails to enter into a sub-contract when it would otherwise be possible he is in breach of the main contract. This may also occur even where the contractor makes an objection because the objection may not be considered reasonable. The supervising officer should therefore respond as quickly as possible to the contractor's objection for failure to do so would cause problems in assessing any damages that flowed from the contractor's breach. The greatest damage flowing from such a breach is most likely to occur when the named sub-contractor withdraws his offer, which he is perfectly entitled to do until it is accepted.

Changing the named sub-contractor

Where a sub-contractor is named in an instruction it is very unlikely that the supervising officer on his own initiative would wish to change his mind. This is possible when pre-contract naming, but no specific provision is made when post-contract naming and one is left to believe that the supervising officer is probably without the right to do so. In the unlikely event of needing to change the named sub-contractor one would in practice overcome the problem by agreement.

Benefits or otherwise of pre-contract naming

Pre-contract naming has the disadvantage of requiring time before the main contract documents can be provided. This means that the preparation of the tender documents will be dependent upon the return of sub-contract tenders.

The extent of work necessary following receipt of sub-contract tenders depends upon whether the sub-contractors are required to design. If no design input is necessary it is possible to overcome the time disadvantage by ensuring that all that needs to be done to complete the main contract tender documents is to append the successful NAM/T. However, where a design input is required this may not be possible because the work may now need to be measured out in detail. One can of course avoid detailed measurement of the work, thus securing some benefit in time but at the expense of a weak framework in terms of valuation.

In all other ways the pre-contract naming of sub-contractors is from the employer's point of view preferable to post-contract naming. This is because in pre-contract naming:

(1) a high level of competition is maintained.
(2) it is not necessary to provide the contractor with a reasonable right of objection to the named sub-contractor, thus avoiding the potential problems of having to rename at a late stage.
(3) the risks associated with the total failure of the named sub-contractor fall more heavily upon the contractor.

The contractor on the other hand benefits from post-contract naming but he does suffer the disadvantage of not knowing who his sub-contractors are at tender stage. Contractors have contended that with a system of nomination such as operated under JCT 80 it is beneficial to know who the sub-contractors are before entering into the main contract because this aids planning. However, in exchange for this information (pre-contract naming) under IFC a price has to be paid by way of the additional risks that are placed upon contractors. Therefore, it is possible that they will now consider this to be to their overall disadvantage and prefer post-contract naming.

Sub-contractors will no doubt view the situation phlegmatically in the belief that either way it matters not unless it is they who are actually named. However, it would appear that a sub-contractor may also benefit from post contract naming whereby the following advantages may accrue:

(1)   the identity of the main contractor is known when tendering.
(2)   the time between the tender date and execution of work on site is likely to be shortened, thus reducing the risks when tendering.
(3)   the later the naming takes place, the more likely it is that a sub-contract will materialise; there is less chance of the supervising officer changing his mind.
(4)   the timing of naming may enhance the sub-contractor's negotiating position in any matters subsequent to tender and prior to contract.

Chapter 3

Form of Tender and Agreement NAM/T

Use of form

The Form of Tender and Agreement NAM/T was produced by the Joint Contracts Tribunal for use where a person is to be named by or on behalf of the building employer as a sub-contractor under the JCT Intermediate Form of Building Contract 1984 Edition in accordance with the first recital or clause 3.3. The purpose of such a form is to standardise the tenders received from sub-contractors and to ensure that a common base covering all the relevant issues is established.

It is intended that NAM/T should be used regardless of whether one is pre-contract naming or post-contract naming a sub-contractor. It has been suggested by some observers that tenderers can be asked to submit tenders in some other way, with only the successful tenderer being required to complete NAM/T. Their reasoning is that although the use of NAM/T is mandatory for creating a contract, it is not mandatory for obtaining tenders. This observation is sound as far as it goes but it does seem rather pointless to obtain tenders by one method and then convert those tenders in order to complete NAM/T, which must be used.

One cannot escape the fact that when naming a sub-contractor, whether pre or post contract, tenders should be sought using NAM/T. To do otherwise will only lead to the problems so frequently encountered with nominated sub-contractors prior to JCT 80.

The major difficulty is that if some other approach is used it will leave the sub-contractor free to negotiate a new price based upon NAM/T. One could argue two possible defences to this point: firstly, the situation is little different to the sub-contractor submitting a tender on NAM/T, then withdrawing it and submitting another tender and secondly, although the approach to obtaining tenders may be different the basis upon which the tender is made can be identical. Neither of these defences is sound. If a sub-contractor withdraws his tender there is every likelihood that he would not be permitted to re-tender, whereas if one asks a sub-contractor to enter into NAM/T having previously invited a tender by some other method you are positively handing the sub-contractor the opportunity to reconsider his price. If the approach is to be different but the basis upon which the tender is to be made remains the same one should not need to depart from the approach intended. Why change from the standard if no real advantage is to be gained?

The composition of NAM/T is most interesting because it is unlike that of other forms used for securing sub-contract tenders. NAM/T comprises three separate sections:

- Section I – Invitation to Tender
- Section II – Tender by Sub-Contractor
- Section III – Articles of Agreement

Therefore, the completion of this form by itself enables a building contract to be created because the form provides for both the making of an offer and its acceptance. The conditions of sub-contract upon which this contract is based are known as NAM/SC. These conditions are referred to in NAM/T and incorporated into the contract by reference. The conditions within NAM/SC are legally enforceable in spite of the fact that they are not signed and regardless of whether they have been read by the parties. Clearly, the contract is intended to be formed by both parties completing the Articles of Agreement, and the NAM/T and NAM/SC have to be taken together when interpreting the contract.

It is most unlikely that a sub-contract will be brought into existence without completion of the Articles of Agreement. This is because the sub-contractor's tender prescribes that the main contractor accepts the offer by completing Section III.

Nevertheless, it is still possible although extremely unlikely that a sub-contract may be created without completion of the Articles of Agreement. However, the main contractor and sub-contractor would be ill advised to do this and should refrain from creating any documentation or act of performance which may be construed as creating or recognising a contract. This is because the terms of such a contract and in some cases the very existence of the contract itself may well become the subject of early litigation.

Having issued such a warning it is interesting to note that NSC/1 (used under JCT 80 for securing nominated sub-contractors by the basic method) does not contain the Articles of Agreement and yet in spite of this a binding contract is formed before they are signed. The Articles of Agreement are together with NSC/4 sub-contract conditions, and this form is adopted by reference once the main contractor signs on the face page of NSC/1 to accept the sub-contractor's offer subject only to the issue by the architect of NSC/3. The issue of NSC/3 brings into operation the terms of the sub-contract and the completion of the Articles of Agreement is simply the formalisation of an already existing contract.

Named sub-contractors are generally considered as domestic sub-contractors and therefore the Joint Contracts Tribunal thought it appropriate that they should use DOM/1 (the standard form of sub-contract for domestic sub-contractors under JCT 80) as a basis for NAM/SC. DOM/1 was produced by the National Federation of Building Trades Employers (now The Building Employers Confederation, BEC) and two specialist sub-contracting associations, so perhaps it is not surprising that it differs in a number of respects from NAM/SC. One significant difference is that DOM/1 is in two parts with one of those parts containing the Articles of Agreement and a fourteen-part appendix. The appendix covers mainly those matters that are contained within NAM/T.

The problem of bringing together a sub-contractor and main contractor is basically much the same in legal terms and it is therefore unfortunate that we have a variety of approaches to achieve the same end result, that of a valid sub-contract. Such diversity without justification can only lead to confusion and one is left believing there is very little justification for these separate approaches.

For the persons advising the contracting parties and the contracting parties themselves the message is clear: ensure that one fully appreciates the point at which a contract is made.

Going out to tender on NAM/T

The supervising officer is required to complete section I, Invitation to Tender, and send a copy to each firm which is to tender for the sub-contract works. If the sub-contractor is to design part or the whole of the sub-contract works the supervising officer should consider using the RIBA/CASEC Form of Employer/Specialist Agreement – ESA/1, discussed in detail in chapter 4.

Where ESA/1 is to be used it should be sent to the sub-contractors tendering for the work. However, the headnote to NAM/T requires that such a separate agreement between the employer and specialist should not be attached to the Form of Tender and Agreement, either where the form is included in the main contract documents or where it is included in an instruction of the supervising officer in respect of the expenditure of a provisional sum.

The reason for ensuring the separation in this way is not clear. One can see that such an agreement is not material to the main contractor in that the main contractor will not be responsible for the named sub-contractor's design (see clause 3.3.7, IFC 84) regardless of whether ESA/1 has been entered into or not and therefore it is not relevant to include it with NAM/T. It may be that the building employer does not wish the main contractor to know whether such an agreement exists. This seems debatable, however, particularly when one considers the position under JCT 80. Here, the main contractor is aware of the use of NSC/2 or NSC/2a and where an alternative method is proposed and it is not the intention to use NSC/2a the contractor must be told. In other words, the reverse situation to that which exists under the Intermediate Form would apply.

Completion of NAM/T by supervising officer

Generally

The supervising officer is required to complete the whole of section I. Much of the detail required in this section is straightforward and needs little if any comment. Therefore, the text of this book confines itself to commenting only upon those points which may give rise to problems or which raise issues of interest.

Sub-contractor's invitation to tender

The sub-contractor should note that he is being invited to tender as a person to be employed by the contractor under a sub-contract in accordance with the Form of Agreement in section III and that his tender should be exclusive of value added tax.

It is worth noting that if a tender were sought on some other basis, as previously referred to, a tender might be obtained with no reference to value added tax. A tender so obtained would be deemed to include value added tax and any sub-contractor quoting on some other basis than NAM/T should specifically qualify that his tender is exclusive of value added tax. It has been decided in the case of *Franks & Collingwood* v. *Gates* (1983) that where a tender is made to a private individual it should be inclusive of value added tax. It is therefore prudent to take account of this decision even when submitting tenders to companies, notwithstanding the fact that the application of this decision may not extend beyond private individuals.

The sub-contractor is required to make his tender by completing section II. His offer must not vary any matters referred to in section I and his tender should be returned to the person specified on page one of NAM/T.

The documents which are sent out with NAM/T and upon which the sub-contractor is required to base his tender are known as 'the numbered documents'. Any such documents, which may include layout drawings, detail drawings, a materials and workmanship specification, performance specification, schedule of works, bills of quantities, format for a sub-contract sum analysis and schedule of rates, shall be listed in item (c) of NAM/T. The documents listed should be clearly numbered for identification purposes to ensure that any subsequent amendments can be easily identified.

Sub-contractor's price build-up

When the sub-contractor makes his tender he is not specifically required to provide a price build-up. Item (c) of NAM/T states:

> 'Should it be decided to name you as the person to execute the sub-contract Works, you will be required to provide the following, hereinafter called "the Priced Documents".'

It is evident from this that it is not the intention to obtain a price build-up before one has decided who one is going to name as the sub-contractor. This is a very strange situation indeed because there are surely many instances when an assessment of this price build-up will be an essential part of the tender evaluation. Relying strictly upon the wording in item (c), if the supervising officer requested a price document the sub-contractor may assume he is to be named as the person to execute the works. This may lead the sub-contractor to believe that he has the contract for the sub-contract work. Although it is extremely unlikely that this alone would create a contract for the sub-contract work it is distinctly possible that the costs of providing the required build-up would be recoverable in the event that the sub-contractor did not secure a contract for the tendered work.

The cost of providing a tender is normally not recoverable and is something which the tenderers themselves have to bear. However, there are exceptions – for instance, where the sub-contractor is required to provide greater detail than would generally be considered usual or where there is an express or implied promise to pay for such work. The situation under NAM/T does not require more information than is usual but it does suggest the possibility of an implied promise to pay for any price breakdown required. This is a possibility, albeit a remote one, because the sub-contractor, relying upon the words used in NAM/T, provides the build-up in the belief that he will secure the sub-contract.

Support for this view can be found in *William Lacey (Hounslow) Ltd* v. *Davis* (1957) where it was held that, although there was no binding contract, there was an implied promise to pay for the tendering service given to the client at his request. Whether or not this view would prevail is open to question because the sub-contractor would normally provide such a build-up without charge. However, the way the tender document has been constructed creates the possibility that the sub-contractor has a case for recovering his costs in providing a price build-up. Such a doubt should not exist.

It is particularly interesting to consider the situation where the supervising officer requires the price build-up of more than one sub-contractor in order to evaluate their tenders. Here it is quite clear that only one sub-contractor will be successful and that there is a possibility that the unsuccessful tenderers will seek some recompense. To avoid this possibility the invitation to tender should be amended accordingly. This can be done by deleting the existing reference on page one of NAM/T which reads 'should it be decided to name you as the person to execute the Sub-Contract Works' and amending the remainder to read 'you will be required to provide the following, hereinafter called "the Priced Documents", with your tender', or if one feels this imposes too great a burden then alternatively 'upon request'. The latter is the most reasonable course so long as only those tenderers under serious consideration are requested to provide a price build-up.

Notice of main contractor(s)
Item (g) of the invitation to tender requires the supervising officer to name the main contractor appointed or, where not so appointed, the main contractors who will be tendering. Where a sub-contractor is to be named after the main contractor has been appointed it is simply a matter of informing the sub-contractor whom, if successful, he will be expected to enter into a sub-contract with. However, if the main contractor has not been appointed at the time sub-contract tenders are sought it is necessary to provide a list of the names of contractors who will be invited to tender.

Item (g) has a side note that says:

'Where item 12(a) applies the Contractor will not have been appointed: instead the names of contractors who will be invited to tender should be sent out.'

This side note is not strictly correct because even where post contract naming is anticipated it is still possible to seek tenders from sub-contractors before the main contractor is known.

The need to provide the tender list of main contractors raises the problem of confidentiality with regard to that list. It is easy to see how the main contractors will establish the names of their competitors. Client organisations differ widely as to their view of this breach of confidentiality. Some take the view that any main contractor worth his salt will find out in any event, but others are most concerned about this issue. They believe that confidentiality should be preserved wherever possible and that its breach should not be assisted by the client himself.

The opportunities for taking real advantage of this knowledge are more limited than one may expect. However, if the client wishes to avoid this breach of confidentiality he must not comply with the requirements of item (g) of NAM/T. This is done by not disclosing the names of the main contractors, but this has the very significant disadvantage that the offer is being made to an unknown firm. If this is what is sought one must for the sake of equity provide for a right of objection once the main contractor is known. To give such a right, however, defeats much of the purpose of pre-contract naming and can cause significant time and design problems.

The client organisation can possibly obtain from the sub-contractors tendering a warranty not to disclose the names of the main contractors. Although this may discourage such disclosure it would not guarantee that it did not occur, nor would it be easy to prove who committed any breach that became apparent. A further possibility would be to give a list of contractors from which the tender list would be drawn.

The purpose of providing the name(s) of the main contractor is to give the sub-contractor the opportunity to decline to tender or alternatively to object to any names on the list of tenders. In reality, few such occasions will arise and therefore the idea of providing a wider list than is to be used is a useful compromise.

Where a sub-contractor does object to a main contractor who has already been appointed then it is clear that any tender from the sub-contractor could not be accepted whilst the objection remains. However, if the sub-contractor objects to one or more of the main contractors on a list it will often be difficult for the supervising officer to decide what to do. In all probability he will allow the sub-contractor to submit a tender but one which could not be accepted by the main contractor to whom he had objected, the logic being that it is not a problem unless that sub-contractor has submitted the lowest tender and the main contractor to whom he has objected has been awarded the contract. However, if a number of sub-contractors objected to a particular main contractor it is likely that the contractor would be dropped from the tender list. This would be an unfortunate state of affairs, for the sub-contractor who secures the naming may not have objected to the main contractor who was dropped from the tender list. Thus the contractor is deprived of the opportunity to secure work.

Notwithstanding the fact that the naming may be made by way of a supervising officer's instruction, the sub-contract tenders can still be sought before the main contractor is known. Where this is to be the case the sub-contractor should be advised of the names of the main contractors tendering. The failure to disclose the names of the main contractors in this situation may cause the sub-contractor to object to the main contractor once known. A reasonable right of objection by the sub-contractor need not be given in the same way as a main contractor's right of objection to a post contract named sub-contractor because the sub-contractor has the right to withdraw his offer so long as it has not been accepted. However, if he is to tender without knowing the names of the main contractors the sub-contractor should make his tender conditional, otherwise he may find it is accepted before he can withdraw.

Main contract information

Information in respect of the main contract is to be provided in items I and II of the invitation to tender. The intention is to provide the sub-contractor with all the relevant information in order that he may assess his bid as accurately as possible. However, it should be noted that this information is in some instances of little effect contractually because it has no direct bearing upon the sub-contract. It is also difficult to appreciate whether the information is intended to comprise a part of the sub-contract itself or simply to be information.

For example, if the access stated was unavailable because the employer was not able to provide it, is the contractor in breach of his sub-contract to the sub-contractor? One submits that he is not because the reference to access is not contained as a contract term.

It is, however, a representation, but the sub-contractor should note that it is not a representation made by the other contracting party – that is, the contractor – and therefore his normal remedies for such a representation which is false are not available. Generally, if a contracting party is induced to enter into a contract on account of a representation made by the other contracting party during negotiations he is entitled to rescind the contract, if certain conditions are met, and may also claim damages.

A distinction is made between innocent and fraudulent representations and this will have an impact upon whether damages are recoverable in addition to the right to rescind the contract. Where a misrepresentation was innocent it would not entitle the other party to recover damages unless the misrepresentation was also a breach of contract. However, it should be noted that under the Misrepresentation Act 1967 the court may in its discretion award damages in lieu of rescission. But where the misrepresentation is fraudulent and loss is suffered, the party suffering damage on account of the fraud may sue for damages in tort in addition to his remedy of rescission.

Therefore, what is the position of a sub-contractor who relies upon and is induced to enter into a contract because of information contained in section one of NAM/T? For example, what remedy has he if the location or type of access is wrongly stated and of material effect? Clearly, there is no misrepresentation by the contractor but there is a misrepresentation made by the supervising officer. If the supervising officer was acting as agent for one of the contracting parties the general remedies, previously described, would be available and not affected. However, in this situation it cannot be said that the supervising officer is acting as agent for either the contractors or the sub-contractors. The supervising officer, however, may well be sued in tort on account of a fraudulent misrepresentation. It is therefore clear that care should be taken to ensure that the information stated is correct and as such both good management and sound in law.

Main contract conditions

The alternative or optional main contract provisions have to be indicated. However, the adoption of bills of quantities on the main contract, for example, does not mean that bills of quantities will be used for the sub-contract. Nevertheless, compatibility is desirable and has been fully discussed in chapter 2.

Any changes from the printed Form of Main Contract Conditions are to be identified and listed. These may not affect the sub-contractor, but if they have any bearing upon him particular care should be given in stating such changes accurately. An example of a change that would be significant in this respect would be an alteration to clause 2.10, defects liability.

Execution of main contract

The supervising officer is required to indicate how the main contract is to be executed – that is, whether it is under hand or under seal. The supervising officer should consult his client before deciding which alternative is appropriate in respect of the main contract. The chief consideration with regard to choice is the limitation period. The limitation period is prescribed by the Limitation Act 1980 and governs the time period in which the parties can sue upon the contract and secure a remedy. The time period for a simple contract (that is, one under hand) is six years and for a speciality contract (one under seal) it is twelve years. The time period runs from the time when the cause of action accrues. This is not necessarily the time when the defect or default is made but may well be when the project is complete.

The sub-contract should be executed in the same manner as the main contract: if the main contract is under hand the sub-contract should be under hand, and if the main contract is under seal so should the sub-contract be. This is required by item 13 in section 1 unless it is stated otherwise. This point is discussed further on page 41. Clearly, it is advantageous that the main contract and sub-contract are subject to the same period of limitation otherwise it would be possible, for example, for the main contractor to be successfully sued where his contract is under seal but for the sub-contractor to plead that the main contractor's claims are statute barred where the sub-contract is under hand. Main contractors should always ensure that they are protected from this situation and anything which purports otherwise should be speedily addressed before they become committed.

There seems to be no reason why the form of execution of the sub-contract should be any different to that of the main contract except possibly when the sub-contract is very small scale. But even then it is debatable.

Inspection of main contract

It is necessary to state where the main contract may be inspected, but item 5 of NAM/T also recognises that this documentation may not be available where pre-contract naming of a sub-contractor is being sought. To be precise, the whole of the contract documentation can never be available if pre-contract naming is required because the naming itself will form part of the main contract documents. Therefore, this documentation will be available when post-contract naming but not when pre-contract naming, and one is left wondering about the philosophy behind this item. Surely if such information is necessary in order that the sub-contractor may accurately assess his tender then it is clear that the sub-contractor who is named at pre-contract stage may well wish to revise his tender once he is able to view the complete main contract documents. Alternatively, it can be argued that such an inspection is a worthless exercise and this being so, why bother to insert where such documents can be inspected?

Main contract appendix

The appendix to the main contract conditions is set out in exactly the same terms in item 6 and is to be completed accordingly. If the naming is post contract the entries made in the appendix will be those which are contained in the main contract appendix, whereas if naming is pre-contract the entries made will be those which it is intended to insert in the main contract appendix. The latter situation gives rise to a problem in that the main contract appendix could be completed differently to how it is set out in item 6. This could occur through error or because of a change of mind. If it occurs because of a change of mind the supervising officer should bring any change to the notice of the sub-contractor before permitting him to enter into a sub-contract with the main contractor.

In practice any change required should be made and the consequences established before the main contract is executed. If a difference occurs because of error and this error is relevant then a significant problem arises, particularly for the supervising officer. It is submitted that the information contained in item 6 will be the basis of the sub-contract and that the new terms actually inserted in the main contract appendix will be irrelevant.

Errors may occur because one is required to address the question of the main contract appendix in the context of NAM/T earlier than one might in the context of the main contract itself. In other words, someone else may decide this information in complete isolation from the person responsible for preparing the main contract appendix. This should be avoided.

Much of the information contained in the reference to the main contract appendix will not affect the sub-contractor, but deciding what does and what does not may cause further problems. The sub-contractor should, however, note that the reference in the appendix to fluctuations is purely for information and does not govern his own recovery of fluctuations under the sub-contract. This is dealt with under item 4 of section II.

Obligations and restrictions

If the employer wishes to impose obligations or restrictions upon those carrying out the works he may do so, but the way in which he does so could be significant in law. Obligations and restrictions may be contained in the main contract conditions and this is the proper way of ensuring that they have effect upon the main contractor. The reason for this is that clause 1.3 (IFC 84) reads:

> 'Nothing contained in the Specification/Schedules of Work/Contract Bills shall override or modify the application or interpretation of that which is contained in the Articles, Conditions, Supplemental Conditions or Appendix.'

This means that if a clause is, for example, written into the specification it can be ignored if it overrides or modifies the main contract conditions. The most likely area where this may occur is in the introduction of further obligations and restrictions.

In effect clause 1.3 reverses the normal legal rule of construction and ensures that the printed word prevails over the written word. This position was accepted, but not unanimously, in *English Industrial Estates Corporation* v. *George Wimpey & Co. Ltd* (1972). However, it should be recognised that simply because an obligation is introduced in the specification it will not always be void on account of being taken to override or modify the contract. This case clearly illustrates that a degree of flexibility and tolerance will be permitted and cognisance will be taken of the written word if it possibly can notwithstanding clauses such as clause 1.3 (IFC 84).

If any obligations or restrictions are incorporated into the main contract conditions then according to item 7 of section I nothing else is necessary to bring it to the attention of the sub-contractor, save that it will be a change to the printed form of main contract conditions and these need to be identified in item 3. Where this occurs and the obligation or restriction is intended to apply equally to the sub-contractor it raises once again the question of whether the reference is merely information or a contract term. In this instance it is necessary for the sub-contractor to observe, perform and comply with the provision, (see clause 6.1.1 (NAM/SC) in so far as it relates and applies to the sub-contractor. The question then becomes: does the provision apply to the sub-contractor?

Where obligations or restrictions are contained other than in the main contract conditions it is necessary, unless they are covered by the numbered documents, that they are repeated in item 7 or that a copy of the relevant part of the main contract specification, bills of quantities or whatever is appended to NAM/T. The presumption is that these references will be included in the numbered documents and only exceptionally will they need to be dealt with under this item. However, the choice as to whether the reference is included in the numbered documents or in item 7 could be significant, as is fully discussed on pages 74–77.

The employer's requirements as to the order of the main contract works is simply another obligation or restriction, but it is necessary to state any such provisions in item 8. A supervising officer should refrain from inserting his own requirements as to the order of works and should do so only if the employer has specific requirements that necessitate it. It is clear that the main contract works include the sub-contract work. Therefore, any reference as to order of works may affect the sub-contract and the sub-contractor is taken to have notice of this. The sub-contractor should pay particular attention to such clauses as they may have a very important impact upon his own programme and tender price.

Location and type of access

The location and type of access need to be stated and presumably in sufficient detail to enable the sub-contractor to locate the access and establish whether it will present any difficulties to him in the execution of his work. Obviously, such matters as availablility and security of access should be prescribed. The failure of the employer to provide the access as described has been discussed on page 33. If the main contractor failed to provide the access described because of his own actions then this is a matter for which the sub-contractor would need to pursue the main contractor. There is little doubt that the access described would be taken to be relevant to the sub-contractor and failure caused by the main contractor to provide it as described would amount to a breach of contract. However, if it was not construed this way then the main contractor would only need to provide reasonable access which could differ from that described.

Changes to the main contract appendix

Item 10 in section I makes provision for inserting revised dates for possession and/or completion of the works where the dates stated in the reproduced main contract appendix in item 6 have been changed. This change of dates is anticipated when the supervising officer is naming by way of an instruction:

(1)   the expenditure of a provisional sum – post contract naming.
(2)   the name of the replacement sub-contractor.

Reference in item 10 to a change of dates should only occur on account of a change in dates that has taken place since the main contract was entered into. Any changes prior to execution of the main contract would, or to be precise should, have been made in the main contract appendix and repeated in item 6 of section I of NAM/T. Therefore, this change of dates referred to in item 10 could come about as a consequence of deferring possession as provided for in clause 2.2 (IFC 84) which in turn provides for an extended completion date, or on account of an extension of time which has been granted to the main contractor.

Other relevant information

Any information which is relevant and relates to the main contract should be given in item 11 of section I. One cannot be sure what matters will fall under this heading, but whatever they are it is perhaps unfortunate that it refers to information relating to the main contract.

The main contract includes the sub-contract and care should be taken to include here for all such matters, although it seems in practice that few issues will occur that are required to be dealt with under this item. If the matter is important it should be dealt with elsewhere; if it is not important perhaps it should not be recorded at all.

Sub-contract information

Generally
Items 12 to 17 of section I relate specifically to the sub-contract yet the supervising officer is required to complete this information. This illustrates that although the contract between the main contractor and sub-contractor is considered to be domestic in nature it is affected in its creation by a third party. The supervising officer is not concerned with the terms of the sub-contract once it has been entered into but, as has been seen, he may be responsible directly to the sub-contractor either personally or as agent for the employer when making a representation in NAM/T, see page 34.

The supervising officer is required to inform the sub-contract tenderer whether the NAM/T, the price document and other numbered documents are to be:

(1) included in the main contract documents.
(2) included in an instruction as to the expenditure of a provisional sum.
(3) included in an instruction as to the naming of a replacement sub-contractor.

The only significance of this is to inform the sub-contractor of the stage at which his appointment will take place. This lets the sub-contractor know whether the main contractor will have a right of reasonable objection to his naming. The right of objection is provided for in clause 3.3.2(c) (IFC 84) and referred to in item 13.

Form of contract

The named sub-contractor is also told in item 13 that he will be required to enter into a sub-contract in accordance with the Form of Agreement in section III with the main contractor selected by the employer. In post contract naming the sub-contractor will be aware of whom the main contractor is but in pre-contract naming he will only know the list from which the employer will select.

As discussed earlier on page 35, the sub-contract should be entered into in the same manner as the main contract unless otherwise stated by the supervising officer. It seems that the supervising officer should not require the sub-contract to be executed in a different manner to the main contract because this will give rise to problems and will provide only very limited advantages in a small number of instances. A sub-contractor tendering for a small amount of sub-contract work may be put off tendering or require a higher price for a contract which is to be executed under seal. However, to be put off tendering is unusual and the additional price to be paid is generally of no real significance.

The question of whether a sub-contract should be entered into under seal when the main contract is executed under hand is easily answered. The answer is no, for it will serve no purpose to the employer. It will allow a remedy for the main contractor against a sub-contractor for a twelve year period but the main contractor can only be liable under the main contract for six years. Therefore, a contractor sued in contract could plead that the claim is statute barred, and where he did he could not successfully sue the sub-contractor because no damage would arise.

The contractor might still be sued by the employer in tort and where he was this might, on account of when time runs in tort, be after the end of the six year limitation period in contract. In these circumstances the main contractor could still sue the sub-contractor under contract. However, it does seem a little senseless to use this process when it is a simple matter to ensure the limitation periods for both the main contract and sub-contract are the same.

Of course the main contractor need not plead the statute of limitations and thus could sue the sub-contractor. This would enable the employer to benefit but no contractor would provide this benefit where there was a cost to himself. It all becomes very complicated, as it can be easily avoided by executing the contracts in a like manner.

Relevance of main contract appendix

This particular part of the sub-contract information is interesting from a legal standpoint as it readily lends itself to differing interpretation. Firstly, one will note that although the reference in item 14 is under the heading of sub-contract information the reference itself is to the main contract appendix. It is provided that 'the main contract appendix and the entries therein will, where relevant, apply to the sub-contract unless otherwise specifically stated here'. This raises two very important questions:

- to what extent are the main contract appendix and the entries therein relevant?
- in what circumstances should matters be specifically stated?

The following indicates which of the items listed are in fact directly relevant to the sub-contract:

(1) whether Article 5.3 applies or not – see reference in NAM/SC, clause 35.3.
(2) whether Article 5.4 applies or not – see reference in NAM/SC, clause 35.4.
(3) extension of time for inability to secure essential labour or goods or materials and deferment of possession – see reference in NAM/SC, clauses 12.7.9, 12.7.10 and 12.7.13.
(4) defects liability period – see reference in NAM/SC, clause 15.3.
(5) the choice of insurance clause – see reference in NAM/SC, clause 9.

It should be noted that if no defects liability period is stated in the main contract appendix the period will be six months, and this applies equally to the sub-contract. This is because item 14 refers to the *appendix* and *entries* where relevant.

The following indicates those items listed which have no direct relevance to the sub-contract:

(1) Article 5.1 – specifically covered in NAM/SC, clause 35.1.
(2) period of interim payments – specifically covered in NAM/SC, clause 19.
(3) fluctuations, clauses 4.9 (a) and C7, 4.9(b) and D1 – specifically covered in NAM/T, section 1, item 16.
(4) insurance cover, clause 6.2.1 – specifically covered in NAM/T section II, item 2.
(5) percentage to cover professional fees.

With regard to the other items listed in the main contract appendix, one cannot be certain as to their relevance. The dates of possession and completion may have a bearing on the sub-contract but because these can be varied for a variety of reasons they do not usually take on any real significance. This issue is further discussed on pages 147–148.

The reference to liquidated damages could well cause confusion as there is, in keeping with other sub-contract contracts, no mention of liquidated damages. It is quite usual for damages in sub-contracts to be unliquidated but here one is left wondering whether the intention is for the liquidated damages specified in the main contract appendix to apply to the sub-contract. It would seem that they are not intended to be applicable and that their only significance is to inform the sub-contractor of the damages that may be applied to a contractor and therefore passed on to the sub-contractor where the sub-contractor is in delay. These damages would only form a part of the claim because the contractor's own damages in addition to these liquidated damages would also be claimed.

The period of final measurement is of no direct consequence to the sub-contract except in so far as it may determine when the final certificate is issued.

The last item in the main contract appendix concerns the date of tender and the relevance of this could easily be debated. However, it is clear that the sub-contract has in clause 1.3 a definition of the date of tender, which reads:

'the date stated in NAM/T, Section 1, item 16 or where not so stated, in Section 11 item 4.'

Therefore, the sub-contract has its own date of tender and the main contract reference is not relevant in the context of fluctuations. However, another reference to the date of tender is also made in section II: 'This tender . . ., is withdrawn if not accepted by the contractor within . . . (weeks) of the date of tender.' This is discussed on pages 49–50.

Once the relevance of the main contract appendix to the sub-contract is understood it is possible to consider in what circumstances one would specifically state in item 14 any differing requirements. Obviously, one should only state specific requirements where they will have relevance, and earlier discussion has identified those matters which may be considered. However, upon further consideration it will be seen that few issues should be considered separately and where they are there may be problems.

The problems arise because the mechanism for stating a changed requirement is not sound. For example, if one wished to omit the equivalent of main contract clause 2.4.10 it would be pointless stating that the appendix reference to clause 2.4.10 should read 'does not apply' because clause 12.7.10 of NAM/SC specifically refers to what is stated in NAM/T, section 1, item 6, and not what is referred to in item 14. This changed requirement could be fulfilled but only by identifying a change to the sub-contract clause itself. This approach would not always be necessary. For example, it is possible to state in item 14 a different defects liability period and a different level of insurance cover. However, the circumstances in which such changes are desirable are difficult to foresee.

Timing of sub-contract works

The supervising officer is required to state the dates between which it is expected that the sub-contract works can be commenced and completed. It is difficult to see how this can have any effect as a contract condition, particularly in the light of the other conditions governing the timing of the sub-contract works. Nevertheless, it is a representation made by the supervising officer, and if this representation was false and the sub-contractor was induced to enter into a contract on account of it then the supervising officer might be at risk (see pages 33–34).

In order to reduce this possibility or the possibility of having to re-tender on account of this period no longer being valid, it is desirable that the supervising officer inserts a generous period during which the sub-contract works can be completed. This naturally has its disadvantages if the period stated is unrealistically long as it will tend to raise the level of the sub-contractor's tender. A very short period may also have this effect but generally a sub-contractor would prefer this situation as it does at least leave some of his options open, i.e. to re-tender or to claim misrepresentation.

Sub-contract fluctuations

The supervising officer is required to complete the relevant parts of item 16 which deal with sub-contract fluctuations and in doing so he must make an important decision with regard to how fluctuations are to be dealt with in the contract.

NAM/SC only provides for two possibilities: clause 33 which provides for fluctuations in contributions, levies and taxes in respect of labour and material, or clause 34 which provides for fluctuations by formula price adjustment.

The adoption of clause 33 is often referred to as a fixed or firm price sub-contract but it is axiomatic that this is not the case. A true firm price would require the deletion of both of these clauses. This is not intended to be an option but it is straightforward enough to achieve if that is what is required. The choice of fluctuations provisions should not necessarily be dictated by the main contract provisions. The sub-contract should be treated on its merits, but the practice of providing wider fluctuation recovery on sub-contracts as compared with the main contract must be seriously questioned. Once the choice has been made either part 1 or part 2 of item 16 is deleted accordingly.

Contributions, levies and taxes

If clause 33 is to be adopted it is necessary to indicate whether duties and taxes on fuels are to be included in the fluctuations adjustment. Fuels are not taken to include electricity as this is specifically referred to in clause 33.2.1. Whether one should include fuels is more a matter of personal preference than of judgment, but it does seem on balance that fuels should not be included unless they are of significant proportions. The situations where fuels are of significant amounts are limited and therefore their inclusion would be limited. However, if one felt it appropriate to adjust for fuels the easiest way is to agree, at contract stage, a notional amount of fuel upon which any variation in tax or levy will be applied.

The supervising officer is also required to state the date of tender. This date is important because it is the base date from which fluctuations are recovered. However, the date stated should not be the date when tenders are to be returned because this means the sub-contractor has difficulty in taking account of any adjustment in levies or taxes which occurs close to his tender submission. It is generally recognised that the date of tender should be about ten days before the submission date and this position is adopted in JCT 80 SFBC where the date of tender is defined as such.

A percentage addition to fluctuation payments has to be stated. Although this is often a nil percentage, 5% is recommended as appropriate. The logic of such an addition is debatable, particularly as it also applies to decreases.

Formula price adjustment

Considering the circumstances in which it is recommended that the Intermediate Form of Building Contract should be used, it seems that the use of formula price adjustment would be the exception rather than the rule. It may, however, be necessary to consider using formula price adjustment where:

(1) there is limited competition available and where as a consequence there is a very strong possibility of sub-contractors either not tendering or submitting very high prices.
(2) the sub-contract tenders are sought well in advance of the main contract tenders.

But even in these circumstances it is only intended as an option when bills of quantities are to be used for the sub-contract work. This requirement will no doubt be ignored in some instances and where it is it will give rise to significant problems in the application of the formula.

The current NAM/SC Formula Rules are those dated September 1984, and it is necessary to indicate which of the methods contained in these rules is to apply to the sub-contract work. This is done by stating whether Part I or Part III (not Part II as referred to on NAM/T in error) is applicable.

Part III is applicable when the tender sought is in respect of specialist engineering work:

- electrical installations
- heating, ventilating and air conditioning installations
- lift installations
- structural steelwork installations
- catering equipment installations

This is because these works have specialist formulae and have their own indices separately published. Sub-contracts for all other work will fall to be adjusted under Part I, which is the work category method and relies upon the separate application of the work category index numbers.

The reference to the non-adjustable element is stated to apply only when the *employer* is a local authority. The intention therefore is that no percentage will be inserted unless the employer under the main contract is a local authority. If, however, a percentage is inserted where the employer is not a local authority then it will be inoperative on account of clause 34.3.3 which states:

'... shall apply to the amount of adjustment under clause 34 provided that this clause 34.3.3 shall only apply where the *Employer* is a Local Authority.' (author's italics)

The employer is defined in clause 1.3 as 'the person with whom the Contractor has entered into the Main Contract'.

NAM/SC formula rules

The references to formula rules will no doubt cause a little difficulty in much the same way as they have with other contracts. This problem is recognised by the side note which states that if certain items are not completed by the supervising officer they are to be completed by the sub-contractor. It may be prudent for the supervising officer to seek the assistance of a quantity surveyor to complete these details. It is preferable that where possible the details are stated and not left to the sub-contractor as each sub-contractor will insert different details, making the evaluation of their tenders that much more difficult.

The disadvantage of doing this is of course obvious in that the sub-contract tenderers may genuinely require different treatment. Deciding for the sub-contractor is therefore to accept a more crude assessment. Nevertheless, on balance it is preferable that the supervising officer does state how these matters are to be dealt with.

With regard to the method of dealing with 'fix only' items, this should be dealt with as follows:

(1) by identifying an appropriate Work Category for the item(s) concerned; or
(2) by stating that it will be included in the balance of adjustable work; or
(3) by the application of a 'fix only' index – to be created in accordance with the Formula Rules.

Where the sub-contract is to be adjusted in accordance with Part I of the Formula Rules it is necessary that a schedule is provided that classifies each item of work under an appropriate heading so that adjustment can be made accordingly. The options available are:

- work category
- fix only
- specialist engineering work (where forming only part of sub-contract work)
- provisional sum work
- balance of adjustable work
- work excluded from adjustment

Where the sub-contract is to be adjusted in accordance with Part III of the Formula Rules it is necessary to insert the weighting of labour and materials for each specialist installation as applicable if more than one is involved under the sub-contract. It is necessary to state the weighting of labour and material because they have separate indices and are therefore adjusted separately.

Where a tender is being sought for a lift installation it is necessary to state when the adjustment of shop fabricated work will take place. It can either be done upon completion of manufacture of all fabricated components or upon delivery to site of all fabricated components. This decision can only be made after considering cash flow, the final cost to be paid and whether the work paid for off site is secure from the employer's point of view (i.e. it cannot be possessed by a liquidator or others).

Attendance

The attendance which the sub-contractor will obtain free of charge from the contractor is set out in clause 25 of NAM/SC. However, it is possible that further attendance may be provided by the contractor to the sub-contractor. Where this is required and known the supervising officer may add further items in item 17 of the sub-contract information, but he may rely upon the sub-contractor to insert any such requirements in section II. In either case once an item of attendance is specified the contractor is required, according to clause 25.3, to provide it free of charge. The contractor does not in fact provide these items of attendance free of charge because he will take account of them when submitting his tender which will include *inter alia* the sub-contract tender and attendance required.

Tender by sub-contractor on NAM/T

The offer

Section II of NAM/T is the sub-contractor's offer and he is required to complete this part of the form as may be necessary. The offer is made to both the employer and the main contractor and is for the sub-contract works identified in the numbered documents referred to in the invitation to tender but subject to the entries made by the sub-contractor himself in section II.

This reference to the entries made by the sub-contractor is extremely important if any of them are unacceptable and the contractor purports to accept the offer but with amendments then a counter-offer is made and no contract is created. This would also apply to entries made under sub-contract fluctuations even though the sub-contractor may not be required to complete the items as they have already been stated in section I.

The price quoted is for a lump sum contract with no alternative provision where complete remeasurement is required, as is the case in Tender NSC/1 – JCT Standard Form of Nominated Sub-Contract Tender and Agreement. The price is also stated as exclusive of value added tax, this being dealt with under clause 17 of NAM/SC.

The sub-contractor offers to conclude a sub-contract with the contractor, either within 21 days of the main contractor entering into the main contract if pre-contract naming is adopted, or immediately upon the issue of a supervising officer's instruction naming the sub-contractor if post contract naming is adopted. It should be noted that the sub-contractor *offers* to carry out the works but does not *undertake* to carry them out. He is under no contractual obligation to do so until a contract is created.

A tender may be withdrawn at any time unless there is an agreement for an offer to remain open for acceptance for a specified period. Any such promise must be supported by some consideration or alternatively made under seal. NAM/T provides:

'This Tender, subject to any extension of the period for its acceptance, is withdrawn if not accepted by the Contractor within ... (weeks) of the date of this Tender.'

However, this is not in itself an agreement to keep the offer open for acceptance and therefore can be withdrawn at any time. But it does indicate how long the offer remains open, and once the time stated has expired the offer has lapsed and cannot be accepted unless the offer is extended or a new offer is made. It may be preferable that the sub-contractor is told how long his tender is to remain open for acceptance. The time runs from the date of tender as stated by the sub-contractor at the end of section II and not from the date of tender as determined for the purpose of calculating fluctuations. This position has to be deduced for it is clear that date of tender is used in two separate ways, yet the date of tender is only specifically defined in the context of fluctuations.

Amongst the matters which the sub-contractor is required to consider when submitting his tender are daywork percentages. Strangely, one has encountered instances where the tenderer has not inserted these percentages, but it is essential that he does so (see Peter R. Hibberd: *Variations in Construction Contracts*, pages 121 to 125). The sub-contractor should note that the percentage on cost he quotes must take account of the contractor's 2.5% cash discount to which he is entitled if payment is made in accordance with clause 19.2 or 19.8.2 as appropriate.

Time requirements

The sub-contractor is required to state the time periods he requires for carrying out the various aspects of his work, having due regard to the dates which have been stated by the supervising officer in item 15 of section I. Such times should accommodate dates so stated but there is no obligation upon the sub-contractor to comply with this. Nevertheless, it is commercial sense to do so where possible. The times stated are relevant and made contractually binding under clause 12 of NAM/SC.

Sub-contract fluctuations

A basic list of materials, goods, electricity and, where relevant, fuels shall be supplied by the sub-contractor where clause 33 is adopted, and attached to his tender. This list identifies those items which then become subject to fluctuations which arise on account of contributions, levies and taxes. As the sub-contractor is usually left to compile this list it is generally in his interests to make it as comprehensive as possible in respect of the significant items included in the sub-contract. Allowing the sub-contractor to compile the list creates a further difficulty in evaluating the tenders, and although not intended it may be that the supervising officer may wish to prescribe those items which are to be subject to adjustment.

If formula price adjustment is adopted the sub-contractor is to identify those articles manufactured outside the United Kingdom and requiring no processing before incorporation into the works which he is specifically required by the sub-contract document to import. In addition the market prices in sterling of such items delivered to site must be indicated on a separate list which is to be attached to his tender. The purpose of identifying these items in this way is because although formula price adjustment is generally applicable to the sub-contract works it is not applicable to such items. Relevant items which are included upon this list are adjusted individually according to actual variations in price.

With regard to the items marked (aa) under item 4 of sub-contract fluctuations, see pages 46–48 for discussion. The remaining items under sub-contract fluctuations are self-explanatory, but once again the supervising officer's attention should be directed to these entries when evaluating the tenders submitted.

Special conditions

The sub-contractor is invited to set out in item 5 any other matters which are relevant to his tender. As the sub-contractor is invited it is tempting for him to think up any matters which he will be able to take advantage of at a later date. However, this desire will be tempered by the fact that it may cause problems when the tender submissions are evaluated. Therefore, the sub-contractor will generally confine himself only to specific matters which need to be referred to.

Tax deduction scheme

Clause 18A (NAM/SC) states that the sub-contactor will either produce his current tax certificate issued to him under section 70 of the Finance Act 1975 or, where the sub-contractor is a user of form 714C, the sub-contractor may as an alternative produce a document as referred to in Regulation 22(i)(c) which is referable to his current tax certificates. Item 6 of section II requires that the sub-contractor identifies which of these he will adopt to satisfy the provisions of this clause.

Submitting the tender

Once the sub-contractor has completed the relevant parts of the tender document he is required to sign and date his offer. The date has significance in that the offer remains open for acceptance from the date stated for the period stated on the first page of section II.

Approving the tender

The supervising officer has to approve the tender and has to indicate whether the tender is included in the main contractor's tender documents or is an instruction; in other words, whether the sub-contractor is being named pre-contract or post contract.

Before the supervising officer approves the tender by counter-signing and dating he should ensure that the time period for acceptance of the offer is sufficient to enable the contractor to accept the offer. If there is any doubt in this respect the supervising officer should agree an extension of the period for acceptance. This may of course entail an adjustment in the tender price and one must remember that the sub-contractor will act to his advantage whenever possible. The supervising officer on the other hand must recognise the difference between a justifiable increase and a situation where the sub-contractor is trying to take advantage. Notwithstanding this the supervising officer still relies upon the sub-contractor to act justly, which in a majority of instances he will do on account of wishing to secure further business.

Any changes to information contained in section I should be notified to the sub-contractor.

Articles of Agreement

NAM/T is both a tender and an agreement and therefore in order to create a binding contract the Articles of Agreement should be completed.

Completion of section III, the Articles of Agreement, is largely straightforward but there are a number of points that require attention in its completion.

Recitals

The first and second recitals require no action but the third recital requires that the numbered documents, the schedule of rates and the contract sum analysis (whichever is appropriate) are signed by both the contractor and the sub-contractor and attached to NAM/T. If the numbered documents have been priced by the sub-contractor these will be the documents which are attached to the NAM/T because the priced copy of the numbered document replaces the unpriced copy.

The fourth recital is concerned with the provisions of the Finance Acts and one is required to indicate the status of the sub-contractor, contractor and employer.

Item (A) requires that the sub-contractor should indicate whether he is, or is not, a user of a current sub-contractor's tax certificate. The recital then goes on to state that where he is then clause 18A of the sub-contract is applicable, whereas if the sub-contractor is not a current user of the tax certificate then clause 18B of the sub-contract shall apply.

Articles

Article one requires, amongst other things, that the sub-contractor shall carry out the works:

> 'in accordance with the requirements, if any, of the Contractor for regulating the due carrying out of the Works which are agreed by the Sub-Contractor ...'

The sub-contractor and contractor may agree any such requirements provided that they shall not alter any item set out in section I or section II of the NAM/T. It seems that such requirements may extend to any matter including that of programme, but it is also apparent that such changes are likely to be small because it is unlikely that the contractor would be prepared to accept a revision in the sub-contractor's tender on account of the change (unless it was a saving). This is because any change falls upon the contractor and cannot be passed on to the employer unless the employer has himself caused the change. If the contractor and sub-contractor do agree to any such requirements they are required to initial the requirements and attach and incorporate them with NAM/T.

The second article states that the sub-contract sum is VAT-exclusive and the parties should take care to ensure that the sum stated does not include value added tax.

Article three makes provision for the name and address of an adjudicator and a trustee-stakeholder to be inserted. The footnote states that the person should be identified at the time of entering the agreement. This is common sense advice, for should a problem arise where their services are required one is not faced with having to agree who should be appointed. Once a problem occurs such agreement is generally very much more difficult to achieve. The purpose of appointing an adjudicator and trustee-stakeholder is to ensure that should a problem arise with regard to any amount the contractor wishes to set-off there is immediately available someone to implement the mechanism provided for in clause 22.

The fee of the adjudicator is initially required to be paid by the sub-contractor as stated in clause 22.8. However, it is apparent that the fee referred to is that which the adjudicator charges if he is required to act. It seems that it would not be unreasonable for an adjudicator to require a retaining fee, bearing in mind that when a problem arises he is to make himself readily available to resolve the dispute. Where such a fee is agreed it seems equitable that it should be shared between the contractor and sub-contractor.

Attestation clause

The agreement should be executed either under hand or under seal as referred to in item 4 and 13 of section I. Generally, the sub-contract will be executed in a similar way to the main contract, and how this is to be done is stated in item 4 of section I.

If the contract is to be under hand it is signed by or on behalf of the contractor and sub-contractor, with each signature being witnessed. If on the other hand the contract is to be under seal, the common seal of the contractor and sub-contractor are separately affixed and witnessed. Where a company does not possess a company seal the contract can still be executed under seal by stating that the contract is signed, sealed and delivered.

Chapter 4

Form ESA/1

Generally

Form ESA/1 is a Form of Employer/Specialist Agreement and was prepared by the Royal Institute of British Architects and the Committee of Associations of Specialist Engineering Contractors. ESA/1 was published in 1984 and is for use between the employer and specialist to be named under the JCT Intermediate Form. The employer/specialist agreement creates privity of contract between these two parties in a way similar to NSC/2 and NSC/2a which are used in conjunction with JCT 80. However, it should be noted that the content of ESA/1 and NSC/2 or 2a is a very different and reflects the fact that a named sub-contractor is not the same as a nominated sub-contract.

Use of form

Where a named sub-contractor is to be used under the IFC 84, regardless of whether the naming is pre-contract or post-contract, it is necessary to consider the use of ESA/1. This form should be used where the named sub-contractor is responsible for any design of the sub-contract works. Design should be taken in its widest sense in that it should be considered to embrace the selection of materials and the satisfaction of any performance requirements.

Where there is some design input on the part of the named sub-contractor it is necessary to adopt the use of this form in order to create a contractual remedy for the employer in the event of any default on the part of the named sub-contractor. The employer will be unable, in the event of a default on the part of the named sub-contractor, to secure a contractual remedy under the main contract against the sub-contractor because there is no privity of contract between them. Furthermore, no remedy exists under the main contract against the main contractor because the contractor's liability in this respect is expressly excluded. Clause 3.3.7 provides that whether or not the named sub-contractor is responsible for design, the contractor shall not be responsible to the employer under this contract for design faults in respect of the sub-contract works.

This means the contractor is free of liability in respect of the named sub-contractor's default in design. This being so it is important that the employer establishes a right directly against the named sub-contractor, and this he does by creating a contract on the terms of ESA/1. It is possible that in any event the employer has a remedy in tort, as following the case of *Junior Books Ltd* v. *Veitchi Co. Ltd* (1982) it was established that a nominated sub-contractor has a duty of care to the employer to see that the work he performs is satisfactory. One cannot be certain that the same would be said of named sub-contractors as they can be distinguished from nominated sub-contractors. Nevertheless, one is left with the impression that the courts would treat them the same, despite the fact that named sub-contractors are domestic in nature, because the named sub-contractor is selected and named by the employer's supervising officer and that seems to be the deciding factor.

Because of the doubt concerning whether a duty exists and the benefits that accrue in having a clearly defined contract, the use of ESA/1 in the circumstances described above is advisable.

ESA/1 is designed to be used in two separate situations:

(1)   when sufficient information is available for a final tender to be produced by the specialist.
(2)   when sufficient information is not available to provide a final tender by the specialist but where an approximate estimate is nevertheless required.

These two situations are referred to in the Form ESA/1 as Procedure A and Procedure B respectively. The test for establishing whether one has sufficient information to proceed to obtain a final tender is whether section I of NAM/T can be completed. If it can, Procedure A is applicable; if not, Procedure B is applicable.

Generally, sub-contracts will be delayed until all the relevant information is available in order that the specialist may submit a firm tender. However, there will be circumstances where this is not possible because the progress of the main contract design is dependent upon the solution adopted by the specialist. In these circumstances Procedure B will be used to obtain an approximate estimate.

It is also evident that the obtaining of approximate estimates once the project has commenced is generally inappropriate because the use of this procedure suggests that the scheme will be redesigned.

There is a major problem in adopting Procedure B because it is necessary at some later stage for the specialist to provide a firm tender. It is evident that if the designer of the main contract works becomes committed because of the nature of the specialist work that has been incorporated then he leaves the employer exposed to the possibility of firm tenders being produced without much regard to the approximate estimate previously given. Obviously, the specialist's reputation may be put at stake, but it is surprising how easy it is to justify the increase that has taken place. Clearly, the designer should be careful to avoid wherever possible putting himself in this position. Unfortunately for his client, it is not always possible to do this when adopting this form of contract procurement.

Completion of form

The supervising officer is required to complete the relevant parts of ESA/1.

Where Procedure A is used it is necessary to delete the following:

- clause B
- the words 'will be submitted by the Specialist to the Architect' in clause 6
- clause BB

and to insert in clause AA the period for acceptance of the sub-contractor's offer.

Where Procedure B is used it is necessary to delete the following:

- clause A
- the words 'is identified on page 3 and annexed hereto' in clause 6
- clause BB

and to insert in clause AA the period for acceptance of the sub-contractor's offer.

If the supervising officer requires the specialist to enter into agreements for the purchase of materials or goods or the fabrication of components for the sub-contract before the sub-contract is executed clauses 7.1 and 7.2 should be left as they stand. If, however, no such provision is required then the clauses should be deleted. It is sometimes necessary for materials to be ordered or work to be fabricated in advance of the execution of the sub-contract. This is due to the need to accommodate long delivery periods and to reduce lead-in times by requiring the specialist to proceed with certain works before the main contractor is known.

The employer should be advised as to whether ESA/1 should be under hand or under seal and his requirements should be indicated by deleting the provision for sealing in the attestation on page 3. How or what is to be done here is not very well expressed, but for further discussion on this point refer to page 73. If the contract is to be executed under seal the consideration of £10 may also be deleted.

Appendix

Parts 1 to 3 of the Appendix are to be completed by the supervising officer and contain sufficient information for the specialist to provide his tender or approximate estimate as required.

When adopting Procedure A it will probably be sufficient to refer in part 1 to the entries made in section I of NAM/T. Where adopting Procedure B, however NAM/T will not be available and the supervising officer should include here as much of the information contained in section I of NAM/T as is possible at the time of seeking the estimate.

When seeking tenders from specialists it may be necessary to provide more information than is currently available. Where this is so the supervising officer is required to identify in Part 2 what this information is and on what date it will be made available.

Part 3 of the Appendix makes provision for the supervising officer to insert his requirements with regard to when the specialist is required to provide certain specified information. This provision is necessary in order that the sub-contract works can be included in the main contract documents or so that an instruction with regard to the sub-contract can be issued as it is essential that such requirements are stated. It is therefore necessary to establish an overall programme for the project in order that such dates can be determined. Provision is made both for information to be supplied as part of the approximate estimate and for information to be supplied with the tender. Therefore, where Procedure B is being followed – that is, where an approximate estimate is required – it will be necessary to insert time requirements under both of these headings to ensure that the sub-contractor is under a continuing obligation should his approximate estimate be proceeded with.

Obtaining offers

Firm tender

Where a firm tender is being sought the supervising officer will send ESA/1 duly completed, together with NAM/T with section I also duly completed, to the specialist firms invited to tender. The tenderers will after completing the relevant parts of NAM/T take a copy. This copy shall be appropriately marked so that it can be seen to relate to ESA/1 and then attached to ESA/1 itself. ESA/1 will be duly completed by the specialist tenderers and then returned together with NAM/T to the supervising officer.

The completion of ESA/1 by the sub-contractor is very straightforward in that he is required simply to insert in AA how the annexed copy of the tender NAM/T has been marked for identification and to sign and date the offer.

Upon receipt of these offers the supervising officer will select the acceptable bid and will arrange for the employer to sign and date ESA/1.

If the agreement is to be under seal it will be necessary that the seals of the respective parties are made on ESA/1. ESA/1 is then retained by the employer with a copy being supplied to the specialist sub-contractor.

Approximate tender

When it is applicable to obtain an approximate estimate it is necessary for the supervising officer to send ESA/1 duly completed to the specialist firms invited to tender. However, in this instance it is not appropriate to send NAM/T because NAM/T is a firm offer and requires that all the relevant information is contained within section I. The specialist will prepare his approximate estimate, which need not be in any particular form unless requested in the invitation to tender. The approximate estimate will then be appropriately marked so that it can be seen to relate to ESA/1 and then attached to ESA/1 itself. ESA/1 will be duly completed by the specialist tenderers and then returned together with the annexed approximate estimate.

The completion of ESA/1 by the sub-contractor is again straightforward but in this case the sub-contractor is required to insert in BB how the annexed approximate estimate has been marked and then to sign and date the offer.

Upon receipt of the approximate estimates the supervising officer will select the acceptable estimate and will arrange for the employer to sign and date ESA/1. The evaluation and selection of an approximate estimate is very much more difficult than that of firm offers. The design team should ensure that an adequate brief is supplied to the specialist in order to assist him to produce his estimates. In due course the supervising officer will complete the relevant parts of NAM/T and forward it to the specialist so that he may submit his tender for the works. In this there is a further difficulty in that one cannot be certain how the firm offer will develop from the approximate tender. This has already been briefly discussed at page 57.

Number of tenders sought

This really depends on the extent and nature of the specialist design input. Where extensive design is undertaken by a specialist tenderer it seems unrealistic to seek, say, eight or ten tenders and perhaps three or four would be more appropriate, the latter number seeking to achieve a balance between ensuring a good measure of competition and wasting resources in the preparation of unsuccessful tenders. If the design input required of the specialist is small then the numbers of tenders sought might be increased to six.

It would not be wise, from the employer's point of view, to unduly restrict the tender list or to adopt a specialist system at approximate tender stage which makes subsequent change difficult because this would have the effect of putting the sub-contractor into a monopolistic position when the firm offer is sought. In other words, a long list of tenderers at approximate estimate stage may only appear to ensure that real competition exists.

Offer and agreement

Clause 1   'The Requirements of the Employer described or referred to herein relate to the Sub-Contract Works. Save to the extent otherwise agreed, the time requirements are to be satisfied by the Specialist, subject to the Employer providing or causing to be provided to the Specialist in the further information (if any) referred to in the Appendix, part 2, at the time or times therein prescribed.'

The time requirements are set out by the supervising officer in part 3 of the Appendix (see page 58). The specialist is obligated to comply with these requirements unless

(1) the employer has failed to provide the further information referred to in part 2 of the Appendix; or
(2) where the parties have otherwise agreed.

Where further information is to be supplied to the specialist after the sub-contractor has received the invitation to tender it is necessary to identify what this information is and by what date it is to be supplied. These details are recorded in part 2 of the Appendix and any failure on the part of the employer to provide such information enables the specialist to escape his liability with regard to the time requirements. This appears to be the position even if the information stated was not required by the specialist in preparing his tender.

Where an approximate estimate is being sought it is possible that because of slippage in the designer's programme the dates stated as time requirements for tender are no longer critical to the designer. Nevertheless, the specialist is still bound by such dates unless the employer has failed to provide the information stated in part 2 of the Appendix.

It is always possible by agreement to amend the dates stated and the reference to 'Save to the extent otherwise agreed' is strictly unnecessary. Any such agreement must be between the specialist and the employer or supervising officer acting on his behalf; it cannot be achieved by the supervising officer acting without the consent of the employer.

Clause 2 'The information required to be provided by the Specialist to the Architect shall also be provided so as to enable the Architect to co-ordinate and integrate the design of the Sub-Contract Works into the design for the Main Contract Works as a whole.'

The information which the specialist provides must enable the co-ordination and integration of this work into the main contract works. The clause must be read in conjunction with clauses 1 and 3 which refer to the time requirements. This means that the information necessary to achieve such co-ordination and integration must be supplied in accordance with the time requirements, notwithstanding the fact that such co-ordination and integration may be achieved without delaying the main contract works when such information is supplied outside the times stated as required.

The extent of information to be supplied must be sufficient to achieve the objective stated in this clause. Unfortunately, problems will arise with regard to what is the appropriate level of information necessary in order to achieve this, and if there is any doubt on the issue it should be clearly defined in the invitation to tender.

Clause 3   'The Specialist is required to provide to the Architect:

3.1   according to such time requirements as are stated in the Appendix, part 3, the information required for the Tender to be used, as the case may be
either
for the purpose of a) obtaining for the Employer tenders for the Main Contract Works and b) inclusion in contract documents of the main contract
or
for the purpose of enabling the Architect to issue an instruction to the main contractor as to the expenditure of a provisional sum requiring the Sub-Contract Works to be executed by the Specialist as a sub-contractor employed by the main contractor

3.2   according to the time requirements stated or referred to in the Tender and/or the Sub-Contract, such further information relating to the Sub-Contract Works as is reasonably necessary to enable the Architect to provide the main contractor with such information as is necessary to enable the main contractor: to carry out and complete the Main Contract Works in accordance with the conditions of the main contract, including the Sub-Contract Works to be executed by the Specialist as the Sub-Contractor; and to provide the Specialist in accordance with the conditions of the Sub-Contract with such information as is reasonably necessary to enable the Specialist to carry out and complete the Sub-Contract Works in accordance with the Sub-Contract.'

Clause 3.1 requires the specialist to provide 'information required for the tender to be used' and this basically means that a Form of Tender NAM/T duly completed by the specialist together with all other relevant information will be made available in accordance with the time schedule. The information to be provided is either:

(1)   to be used in the main contract documents for the purposes of obtaining a tender from the main contractor for the employer (pre-contract naming); or
(2)   to enable the architect to issue an instruction to expend a provisional sum provided in the contract documents (post-contract naming).

It is unclear what level of information is required to be provided in order that this objective may be achieved. However, it does seem that where the specialist is providing some design input it is unlikely that a measured bill will accompany the tender. This means that the main contract documents will generally include an all embracing item for the sub-contract work in spite of the fact that some commentators do not believe that this is the intention. The reality of the situation will most certainly determine that full measurement of this work is the exception rather than the rule.

Under clause 3.2 the specialist is also required to provide any further information necessary to enable the main contractor to carry out the works and also to enable the architect to provide to the specialist himself the information necessary for the sub-contractor to carry out his works. The second part of this clause appears rather contorted but it simply recognises that although the specialist prepares the information this must first be passed to the architect for approval, or whatever, before the specialist can be formally provided by the architect (via the main contractor) with such information necessary for performing the sub-contract. This prevents the sub-contractor from arguing that he is unable to progress the sub-contract because he has not received information from the architect when the reason for his not having received the information is his own default in that he did not provide it in the first place.

Clause 4   'The Employer shall be entitled to use for the purposes of carrying out and completing the Main Contract Works and maintaining or altering the Main Contract Works any drawings or information provided by the Specialist in accordance with this Agreement.'

A specialist who is invited to tender for sub-contract works and who enters into ESA/1 does, it seems, put himself at some unnecessary risk. The purpose of clause 4 is unclear because it may be simply to ensure that the designer can incorporate the specialist's design into his own without problems of copyright. However, it is most unlikely that this problem would materialise when the sub-contractor who has designed the works is also contracted to perform the works. Therefore, one may be left with the view that the clause is provided in order to enable designs of sub-contractors to be adopted even though those sub-contractors are not employed to perform the works.

Is the specialist designer entitled to any recompense if this should happen? This matter needs to be looked at from two angles: when obtaining a firm tender, and when obtaining an approximate estimate.

Firstly, when one is obtaining a firm tender no mention of recompense is made and therefore it is likely that some will believe that there is no entitlement to payment. It is well established law that a tenderer will not be paid for the costs of preparing a tender: see *William Lacey (Hounslow) Ltd* v. *Davis* (1957). However, it was also decided in this case that a claim may arise in quasi-contract for the reasonable cost of providing tenders where the work involved goes beyond that which can be normally expected. Generally, design costs would be considered as part of the tendering process and as such would not be recoverable unless the employer requires detail beyond that which can be normally expected or makes some use of the design.

It is with this last point that we are concerned for if that is the case then the specialist would under common law be entitled to his reasonable costs on a *quantum meruit* basis. However, the position under ESA/1 is probably different for here the specialist is entering into a contract and as a part of that contract is required to provide a tender and other information, knowing that his design may be used and that he may not be required to execute that design. This appears to be a most unlikely situation but nevertheless it is possible that the clause is provided to avoid complications where a specialist becomes unable or incapable of performing the works.

It is possible that the courts would imply a term for payment of a reasonable sum where this occurred, but the contract read as a whole is against this view. This is because it can be shown that in the alternative procedure for obtaining an approximate estimate payment is specifically referred to, and therefore it may be reasonable to deduce that the absence of such a provision when obtaining a firm tender is a clear intention that no payment is to be made. It must be stressed, however, that such a deduction does not provide one with a conclusive evidence as to the intention of parties, and it is conceivable (although, it is submitted, unlikely) that a term may still be implied to the effect that certain costs are recoverable.

Therefore, the specialist who finds himself in this situation is unlikely to secure any compensation. This would not be particularly fair where the sub-contractor wished and was able to perform the works he had designed, and one is left wondering whether the specialist might have an action in fraud: see *Richardson* v. *Silvester* (1873). Unfortunately for the specialist, it does seem he would be most unlikely to be able to prove that such a fraud existed because he had himself agreed to the possibility of such action by the employer.

The position when the specialist is providing an approximate estimate is somewhat different for it provides specifically in item BB for recompense for design costs where a sub-contract is not entered into. The probable philosophy behind this difference is that the design input must generally be of greater significance when adopting this approach because it affects the development of the main contract works and there is a greater risk that such work will not proceed to the stage where a sub-contract is let.

This philosophy is sound as far as it goes but it does not take account of the scale of the design by the specialist. In other words, a specialist giving a firm tender is unable to recover his costs even though it necessitated large scale design work, whereas a specialist giving an approximate estimate is able to recover his costs. An approximate estimate is necessitated because the work under consideration affects the development of the scheme but in reality his design costs may be significantly less. The specialist tenderer can therefore see that he may recover his tendering costs not because of how much design he undertakes but because of when such design is undertaken.

Item BB which refers to the approximate estimate, provides:

'that if the Sub-Contract is not entered into by the Specialist, the Employer shall pay the Specialist the amount of any expenses reasonably and properly incurred by the Specialist in carrying out work in the designing of the Sub-Contract Works in anticipation of the Sub-Contract and in accordance with the Requirements.'

Where the work does not proceed the sub-contractor is entitled only to his expenses properly incurred and is not entitled to any profit from the cost of design work. One must stress that under the contract term the sub-contractor is entitled to expenses only. There is no question that he will be paid the worth (on a quantum valebant) for the work he has done or for any profit in respect of that work. This leaves the specialist open to possible abuse.

From the above it is clear that the specialist is entitled to recover his expenses if a sub-contract is not entered into regardless of whether it is the employer who does not proceed to name the specialist or the specialist himself who does not wish to enter into a sub-contract. It is unlikely that a specialist would not wish to enter into a sub-contract, but this may occur where he is unhappy with the successful main contractor. Clearly, the specialist is not obliged to enter into a sub-contract but he is, however, still bound by the obligations contained in ESA/1. Therefore, a specialist will generally be inclined to enter into a sub-contract notwithstanding any reservations he may have about the main contractor.

If the specialist did decline to enter into a sub-contract he would in accordance with item BB be entitled to his expenses incurred to date and, it seems, his expenses associated with his continuing obligations under ESA/1, for BB refers to expenses incurred ... in anticipation of the sub-contract *and* in accordance with the requirements. A similar situation would prevail if the employer prevented a sub-contract coming into existence because he decided not to name that specialist.

> 'Clause 5   The Requirements include the exercise by the Specialist of reasonable care and skill in:
> – the design of the Sub-Contract Works insofar as the Sub-Contract Works have been or will be designed by the Specialist and
> – the selection of materials and goods for the Sub-Contract Works insofar as such materials and goods have been or will be selected by the Specialist and
> – the satisfaction of any performance specification or requirement included or referred to in the description of the Sub-Contract Works included or annexed to the Tender.'

This clause is apparently straightforward in its intention but it raises three important questions:

(1)   is such a clause likely to stand up in law?
(2)   how in any event does one establish whether reasonable care and skill have been exercised?

(3) where and how is the extent of the sub-contractor's obligations in respect of these matters defined?

The validity of a clause such as this has been questioned. John Parris, at page 111 in the second edition of his book *The Standard Form of Building Contract: JCT 80*, makes reference to clause 2.1 of NSC/2. This clause is similar to the one in question and he comments as follows:

'That is, again an attempt to exclude by implication the implied warranties as to good work and materials and fitness for the purpose.'

John Parris believes the courts would not permit what is clearly a restriction upon the standard that could otherwise be expected.

Many building contracts used in the U.K. exclude a warranty as to fitness for purpose. This is because someone else has assumed the role of designer – that is, for instance, an architect or engineer – and the contractor is not generally expected in these circumstances to take on such a liability. However, the position under a design and build contract is generally considered to be that the contractor does give a warranty as to fitness for purposes. This is notwithstanding the fact that attempts are made in design and build contracts to restrict such liability to the use of reasonable skill and care. The argument that is made is simply that the designer must design for the purposes made known to him or that can be implied in the project, and the exercise of reasonable skill and care will provide a design that is fit for the purpose.

One can see defects in the logic when the argument is applied to the satisfaction of any performance or requirement. Where a performance specification is given the purpose of the works is either generally or specifically defined. Therefore, using the argument expressed above the exercise of reasonable skill and care must satisfy the performance specification.

It is evident that it is possible to exercise reasonable skill and care but still fail to provide the absolute standard required by fitness for purpose. For example, a steel framed building together with its foundations have been designed using all reasonable skill and care but the foundations fail owing to movement of the bearing strata which could not have been reasonably foreseen by the designer.

Although the logic is not sound it does clearly show how a clause such as clause 5 of ESA/1 can be widely construed so as to embrace where relevant fitness for purpose.

Therefore, taking this point together with that expressed by John

Parris, one is led to the conclusion that the courts could either:

(1)   reject the words 'reasonable care and skill' and require a design that is fit for the purpose; or
(2)   interpret the words 'reasonable care and skill' to embrace fitness for purpose.

Either way the outcome would probably be the same. But, it is submitted, the courts would probably hold that the sub-contractor in exercising reasonable care and skill is giving a warranty for fitness for purpose unless in the circumstances it could be implied as unreasonable.

The case of *Greaves (Contractors) Ltd* v. *Baynham Meikle & Partners* (1975) is relevant to this issue. This case shows that the designer knew the building was to be used by fork lift trucks and therefore he was required to design to accommodate *fully laden* fork lift trucks. If the designer failed to do so he would be in breach of his duty to exercise reasonable skill and care.

The above reasoning can be applied to a design which is for the whole works or only part of the works. However, where one is concerned with part design there is the additional problem of the overlap and integration of work designed by different people.

The question of fitness for purpose is made that much more difficult where this occurs and even more complicated by the fact that a contract such as ESA/1 probably falls within the provisions of Part I of the Supply of Goods and Services Act 1982. This Act implies warranties to the effect that:

(1)   the goods will correspond with the description.
(2)   the goods shall be of merchantable quality.
(3)   the goods will be *fit for the purpose required.*

Therefore a sub-contractor as designer must take account of this in the selection of materials for his design.

If the sub-contractor attempts to exclude or restrict his liability in this respect the contract term is subject to the Unfair Contract Terms Act 1977.

The question of whether reasonable care and skill have been used is generally judged against that imposed by the Bolam standard of professional care. This was laid down in the case of *Bolam* v. *Friern Hospital Management Committee* (1957), but discussion on this point is beyond the scope of this book.

It is interesting to note that one generally refers to 'reasonable skill and care' but for some unknown reason clause 5 refers to 'reasonable care and skill'. This will probably have no material impact but it is another illustration of the inconsistencies that occur and which do nothing for one's comprehension of the subject.

Where and how the extent of the sub-contractor's obligations are defined in respect of:

- the design of the sub-contract works
- the selection of materials and goods
- the satisfaction of any performance specification or requirement

is a very important practical problem which may have significant legal consequences.

The first of the problems that is encountered concerns how any obligations with regard to these matters are imposed upon the sub-contractor. The problem can be appreciated by reference to Dennis Turner's book *Building Sub-Contract Forms* where at pages 255 to 257 he discusses this issue albeit in respect of NSC/2 and NSC/4. He commences by stating:

'A wide issue is whereabouts in the documentation the extent and nature of design and other aspects should be delineated. Under the sub-contract, the sub-contractor has no responsibility over these matters because the contractor has none ... so that in this respect the sub-contract documents are not an appropriate vehicle for giving requirements.'

It is submitted that this reference, although identifying a potential problem, is misleading for the following reasons:

(1) the reference to 'under the sub-contract' is, it seems, simply to the sub-contract conditions but the conditions themselves may embrace other documents; clause 4.1.1 of NSC/4 makes reference to sub-contract documents and these are defined in clause 2.1.1. and may include 'any documents annexed thereto or referred to therein'.

(2) where such documents have been incorporated there is nothing in the sub-contract conditions that expressly excludes responsibility for design, etc.

(3) the sub-contractor may have a design responsibility even though

the main contractor does not have one for his or the sub-contractor's work; however, it is recognised that the sub-contractor may be under an obligation under the sub-contract which will not be enforced by the contractor because he has no liability under the main contract and furthermore the employer cannot get redress in contract because there is no privity of contract between the employer and the sub-contractor; that is unless a collateral agreement exists.

The case of *Junior Books Ltd* v. *The Veitchi Co. Ltd* (1982) establishes that an employer may have a remedy in tort against a defaulting nominated sub-contractor. What is contained in the sub-contract between the main contractor and the sub-contractor may therefore have some bearing on whether there is a breach of the duty of care. Nevertheless, to rely on a tortious remedy is not advisable, hence the use of collateral agreements such as NSC/2 and ESA/1.

If a collateral agreement exists and the sub-contractor has taken on a design responsibility then the extent of this should be delineated and specifically expressed in this context, for if it is only expressed in the sub-contract documents this may, under the nomination arrangements for JCT 80, give rise to problems. The difficulty is that the various documents are not necessarily sufficiently tied in together to create a contractual obligation upon which the employer can successfully rely.

Under the naming provisions of IFC 84 and the use of NAM/SC and ESA/1 this situation has been very much improved. Both AA and BB of ESA/1 state:

'We offer to satisfy the Requirements described or referred to in the Schedule above in accordance with this Agreement.'

and with a copy of the tender annexed.

In order for the sub-contractor to submit his tender sub-contract documents must be provided by the design team, and these documents then form the basis of both the offer to the employer (ESA/1 and NAM/T) and the offer to the contractor (NAM/T). The documents must contain, if relevant, the extent to which the sub-contractor is required to design, select materials or goods, or satisfy any performance specification.

Any such references should be clearly expressed and the work required precisely defined because in this respect the sub-contractor is obligated 'insofar' as the sub-contract works have been or will be designed etc. by the specialist. Any references only appear to establish a minimum level of requirement, because should the sub-contractor provide design in addition to these requirements there is nothing in clause 5 to limit his obligations in respect of this further design.

Clause 2.1 of NSC/2 contains the words:

'Nothing in Clause 2.1 shall be construed so as to affect the obligations of the Sub-Contractor under Sub-Contract NSC/4 in regard to the supply under the Sub-Contract of workmanship, materials and goods.'

A similar reference is not made under Clause 5 of ESA/1.

Definitions

Clause 6 contains a list of definitions which are to apply to ESA/1 regardless of whether the specialist's tender is accepted and a sub-contract entered into.

The list of definitions need little explanation or comment, save as follows. Contractors and sub-contractors alike are often more than interested in who the architect or supervising officer will be. It has been suggested that this may even affect the tender price. This may well be justified but the sub-contractor should note that the supervising officer in the definition includes:

'or such other person as the Employer appoints in place of the person so named ...'

This means that the sub-contractor has no right of objection to the substitute supervising officer and is bound by the contracts he has entered into. The supervising officer has no direct say in NAM/SC so a change would have little effect on this contract. However, he does have direct impact on the fulfilment of the sub-contractor's obligations under ESA/1.

The Form of Tender NAM/T is defined but no reference is made to the current edition of this form. At the moment this form has not been revised, but should this occur it would be prudent to amend the definition accordingly. This will ensure that no sub-contractor will refuse to tender on the grounds that the latest edition of NAM/T has not been used. It may also avoid the complications caused by entering into a different edition of the form to the one stated.

Information is stated to include, wherever appropriate, drawings and the information to be submitted with the approximate estimate and/or the tender. Unfortunately, this does not establish precisely what may be required, and the point is discussed under clauses 3.1 and 3.2.

Materials and goods

Clauses 7.1 and 7.2 are intended to be optional (see pages 57–58). It seems that generally these clauses should be left in because although one can reasonably establish the circumstances when the early purchase of materials and goods or fabrication is necessary there may be occasions when it is required but could not be anticipated. Should this latter event occur a separate agreement would need to be entered into, and this may be to the disadvantage of the employer.

If the clauses are left to operate the employer has a clear right to require the specialist to proceed with regard to these matters. The cost of this item would be a part of his tender and the sub-contractor would normally obtain payment under his sub-contract with the main contractor. However, if no sub-contract was entered into payment would not be forthcoming by this means. Therefore, clause 7.2 is intended to regulate the position and provide for the sub-contractor to be paid the cost of such materials, goods, etc., and for the property to pass to the employer upon payment.

Although it is perhaps wise to maintain clause 7.1, the deletion of clause 7.2 is not so important and its deletion does not mean that the sub-contractor is not entitled to payment. In the absence of such a clause or other agreement the sub-contractor would be able to recover the costs of the materials, goods, etc., which have been ordered by the employer on a quantum valebant. This is a similar situation to that which would exist if both clauses 7.1 and 7.2 were deleted and where the sub-contractor supplied materials and goods (without a formal agreement existing) at the request of the employer. The absence of the clause would not generally affect the passing of property in the goods, because to prevent this happening a valid retention of title clause would be required and this is unlikely to be present in the circumstances under consideration.

It should be noted that it is the employer who has the right under clause 7.1 to require the specialist to proceed with the purchase of materials, goods, etc., not the architect or supervising officer. It would generally be assumed that a supervising officer giving such an instruction to proceed would do so as agent for the employer. Nevertheless, it would be advisable for the sub-contractor who is confronted with such a request to ensure that the supervising officer does in fact act as the agent.

Offer and agreement

AA and BB both state:

'This offer is withdrawn if not accepted within … weeks of the date of our signature below.'

The general position concerning the withdrawal of offers was discussed at pages 49–50 in connection with NAM/T and is equally applicable to ESA/1. However, it is possible to construe that the offer must remain open for the period specified and that it cannot be withdrawn without agreement. This construction is possible because of £10 consideration that shall be payable to the specialist. It is not intended that the consideration is given for the purpose of keeping an offer open but rather to ensure that there is some consideration in order that a contract may be created. Unfortunately, this is not particularly clear and an improvement in the drafting is desirable.

Chapter 5

Sub-contract documents – NAM/SC

Generally

The sub-contract conditions are deemed to be incorporated in article 1.2 of the Articles of Agreement in section III of NAM/T. This means that the parties do not need to sign the sub-contract conditions themselves. The sub-contract conditions from part of the sub-contract regardless of whether they have been signed if the offer and acceptance is on the basis of NAM/T. The sub-contract conditions do not contain any specific place for the appending of signatures as they are intended to be incorporated into the contract by reference.

NAM/SC is similar but by no means identical to the DOM/1 sub-contract conditions, providing as it does for the domestic sub-contracting relationship. The use of DOM/1 for domestic sub-contracting when using JCT 80 is not obligatory, whereas the use of NAM/SC is a mandatory part of the procedure for a domestic sub-contract created by the naming provisions under IFC 84.

As NAM/SC is an integral part of the procedure it is befitting that the JCT themselves have produced the sub-contract conditions. This is a departure from previous practice whereby the domestic sub-contracts for use in connection with JCT 63 and JCT 80 have been produced by the NFBTE (now BEC), FASS and CASEC. Nevertheless, the JCT used DOM/1 as a model for the NAM/SC sub-contract conditions and as previously stated they do contain some similarities.

Sub-contract documents

Clause 1.2 makes it clear that the sub-contract is to be read as a whole and the effect or operation of any recital or article in NAM/T or clause in the sub-contract conditions is subject to any relevant qualification or modification in any other part of these documents. The common law position is that contracts should be read as a whole. This requires the court to establish firstly, what documents form the contract and secondly, to give effect to all those terms and to reconcile any inconsistencies between the various documents. Therefore, there is no great significance in the first part because the contract documents are defined in clause 2. However, the reference in clause 1.2 recognises that the NAM/T may have impact on the sub-contract and the sub-contract, unlike some other contracts, does not preclude the intentions of NAM/T.

For example, an item contained in the preliminaries of a bill of quantities when using JCT 80 may not be legally binding. This can occur where an item overrides or modifies the contract conditions because the contract contains clause 2.2.1 which states:

'Nothing contained in the Contract Bills shall override or modify the application or interpretation of that which is contained in the Articles of Agreement, the Conditions or Appendix.'

The effect of this clause has been considered in a number of cases, for instance:

- *Gleeson* v. *Hillingdon LBC* (1970)
- *North West Metropolitan Regional Hospital Board* v. *T A Bickerton & Son Ltd* (1970)
- *English Industrial Estates Corporation* v. *George Wimpey & Co. Ltd* (1972)
- *Bramall & Ogden Ltd* v. *Sheffield City Council* (1983)

Because of the existence of this type of clause in these contracts, the court took the same view on each occasion that the condition cannot be overriden or modified or affected by what is contained in the bills. The position would be the same if the provision were contained in some document other than the bills. Conversely, a preliminary item may have effect so long as it does not override, modify or affect the condition.

The fact that the legal position recognises that where such a clause exists the bills cannot override, modify or affect the sub-contract condition has been the subject of strong criticism because in the words of Lord Justice Stephenson '... it follows from a literal interpretation of clause 12 (JCT 63) that the court must disregard – or even reverse – the ordinary and sensible rules of construction ...'. Lord Denning's minority view in the same case (*English Industrial Estates* v. *Wimpey*) was that type should always prevail over print and that a bill item should prevail over clause 12 (1) of JCT 63, i.e. clause 2.2.1 of JCT 80.

Therefore, perhaps as a consequence of these observations we now see in clause 2.1.2 of NAM/SC a somewhat differently worded clause:

'Nothing contained in any descriptive schedule or other like document issued in connection with and for use in carrying out the Sub-Contract Works shall impose any obligation beyond those imposed by the Sub-Contract Documents.'

The reference to 'documents' as opposed to conditions is of fundamental importance because this opens up the possibility that the legal position may be expressed in a document other than the conditions. Therefore, in accordance with clause 1.2 the sub-contract can now be truly read as a whole because the sub-contract documents are those referred to in article 1, namely:

* tender and agreement NAM/T
* sub-contract conditions
* numbered documents

The question which still exists is what happens now if there is ambiguity or conflict between the various documents? Which document prevails where conflict exists? In the absence of a clause which determines which documents are to prevail, the normal rules of construction would apply and the written word would prevail over the printed. Notwithstanding this, NAM/SC sets out to establish a priority of documents so that where conflict does exist it can be resolved.

Clause 2.2 states which document will prevail in the event of a conflict between the various documents:

(1)   NAM/T will prevail over the sub-contract conditions where there is conflict between these documents.
(2)   NAM/T and sub-contract conditions will prevail over the numbered documents (for example, drawings, priced schedule) where there is conflict between these two sets of documents.
(3)   the sub-contract documents will prevail over the main contract where there is conflict between these documents.

The following illustrates this in a diagrammatic way. The reference to 0 and X represents the item as it appears in the document and the box indicates which item would prevail. The dotted box simply indicates that the numbered documents happen to be the same as the prevailing document.

Numbered Documents	0	0	0
NAM/T	X	0	X
Sub-Contract Conditions	X	X	0

From this it can be seen that NAM/T is the pre-eminent document where there is conflict between the terms. Therefore, it seems we do not move to the common law position where the numbered documents (which would generally be written) would prevail over the other documents which are either entirely or in part printed. Nevertheless, a significant move in that direction is made because the most likely areas of conflict will arise from the written (as opposed to printed) parts of NAM/T and this document will now be seen to prevail. Lord Denning will no doubt be much happier with this situation.

The last part of clause 2.2 whereby the sub-contract documents are said to prevail over the main contract really goes without saying. It is intended to clear up any conflict between the words purportedly extracted from the main conditions and incorporated in the sub-contract and those which are actually written into the main contract conditions.

Notification of discrepancies

There is no express requirement in NAM/SC, as there is in DOM/1 clause 2.3, that the sub-contractor should give notice of any discrepancy or divergence he may find in or between the various sub-contract documents. Nevertheless, it may be prudent for the sub-contractor to indicate to the contractor any discrepancy or inconsistency he finds because it is possible that an implied obligation to notify such inconsistencies may be purported. This would not be inconsistent with the fact that the contract expresses a clear intention as to which document prevails and that the sub-contractor should be entitled to proceed on this basis because this clause is obviously intended to state the position should a discrepancy or inconsistency not be found until after the work in question had been commenced or completed.

To summarise, it would seem right that the sub-contractor upon discovery of a discrepancy should notify the contractor but where he does not do so he may still be protected if he complies with the prevailing document.

However, a different position may exist where one is concerned with the issue of further drawings or details which are reasonably necessary to enable the sub-contractor to carry out his work. The situation would be no different to that expressed above if the inconsistency was between the further drawings and the sub-contract documents. But if the inconsistency was within the further drawing or details it is apparent that the sub-contractor could not act in the knowledge that one document is prevalent to another. In these circumstances he must notify any inconsistency he finds and it would, it is submitted, be an implied term that he did so. If a discrepancy exists and the details indicated would not work, the sub-contractor is clearly under a duty to report to the contractor the design defects, if known or should have been apparent. This was decided in the cases of *Equitable Debenture Assets Corporation Ltd* v. *William Moss and others* (1984), and *Victoria University of Manchester* v. *Hugh Wilson and Lewis Womersley and Pochin (Contractors) Ltd* (1984).

Chapter 6

Control of the works – NAM/SC

Directions of the contractor

Generally

It is generally considered that the supervising officer has no direct control over the work of a domestic sub-contractor. This in essence is little different from a nominated sub-contract because in both situations the sub-contractor is under the direct control of the main contractor by virtue of the sub-contract between them. It is not absolutely true that the supervising officer has no direct control over the sub-contract work because matters may be left for his opinion (see clause 5.1). However, the extent of indirect control is significant and this chapter will, amongst other things, identify how this is effected.

Directions

The sub-contractor in accordance with clause 5.1 is required to conform with all the reasonable directions of the contractor. Note that the word used is not instruction but direction. It seems that the term 'instruction' is reserved solely for the supervising officer under the main contract conditions, with direction used in this instance in order to draw out the distinction. In practice there is little difference between the effect of the two in this context because, as will be seen in clause 5.2.2, any instruction of the supervising officer shall be deemed to be a direction of the contractor.

The following tables show the clauses under which the contractor could issue directions to the sub-contractor.

Table 6.1 Express provisions where the contractor shall issue directions.

Clause	Event that gives rise to direction
2.4	Inconsistency in or between the Sub-Contract Documents or drawings or documents issued under clause 2.3.
3.4	Departure from the method of preparation of bills of quantities.
5.2.2	Written instructions of the Architect/Supervising Officer issued under the Main Contract affecting the Sub-Contract Works.
15.4*	Where Architect/Supervisor instructs that shrinkages or other faults are not to be made good ... then the Contractor shall *instruct* the Sub-Contractor.

*This clause is inconsistent in that the word used in this clause is instruct and not direct.

Table 6.2 Express provision where the contractor is deemed to have given directions.

Clause	Event that gives rise to direction
5.2.2*	Written instruction of the Architect/Supervising Officer issued under the Main Contract affecting the Sub-Contract Works.
16.2.9	Where work is substantially changed as a consequence of a variation

* Note: this clause does not require the contractor to issue a separate direction but he is obliged to pass on any written instruction of the supervising officer because he is himself required to comply with these instructions. An instruction which is forwarded to the sub-contractor is then deemed to be a reasonable direction of the contractor.

Table 6.3 Express provisions where the contractor may issue directions.

Clause	Event upon which a direction may be issued
5.2.1	Any event in regard to the Sub-Contract Works.
5.5	Opening up for inspection or for arranging tests.
5.6.2	Opening up for inspection or for arranging tests of further work where the Sub-Contractor has not complied with clause 5.6.2.
15.3	Making good defects.
16.1*	Variation or expenditure of provisional sum.

*This will normally only be operated in accordance with the supervising officer's written instruction and will then fall to be a direction under clause 5.2.2.

Table 6.4 Express provisions where the contractor shall act and where his actions may give rise to a direction.

Clause	Event which may give rise to direction
10.1	The requirement of documentary evidence that the insurances required are properly maintained.

Scope of directions

It is clear from the foregoing tables that the contractor has the contractual right to issue directions to the sub-contractor over a wide range of matters, and this authority is specifically embraced in clause 5.1:

'The Sub-Contractor shall carry out and complete the Sub-Contract Works in compliance with the Sub-Contract Documents and in conformity with all the reasonable directions of the Contractor (so far as they may apply) regulating for the time being the due carrying out of the Works ...'

This clause amongst other things sets out the sub-contractor's general obligation with regard to compliance with the contractor's directions. The sub-contractor is not obliged to comply with all the directions that a contractor may issue. Firstly, the direction must be one issued under the major contract. Secondly, even if the direction is so provided for in general terms it must still nevertheless be reasonable.

The words 'so far as they may apply' in clause 5.1 are not particularly helpful in the interpretation of this clause, but presumably it can be interpreted to mean:

(1) the sub-contractor is not obliged to comply with a direction that is not relevant to the sub-contractor (this seems obvious); and/or
(2) the sub-contractor is not obliged to comply with a direction for which there is no contractual provision; and/or
(3) the sub-contractor is obliged to comply with an unreasonable direction to the extent that it would be reasonable to do so – in other words, the overall effect of a direction may well be unreasonable but the sub-contractor may be obligated to comply with the direction to the extent that it could be considered reasonable.

The overall effect of the words 'so far as they may apply' seems to embrace all of these points, although not in their entirety. But the main issue of contention is likely to arise if one is called upon to determine what is a reasonable direction. There is no clear-cut answer to this problem and one would have to consider the facts of each situation in order to make a judgment.

From a legal point of view this seldom causes undue concern but there are nevertheless many examples of the problem. If a concrete mix only is specified, for instance, is it reasonable that the sub-contractor should comply with directions that require a particular type of cement or a specific method of mixing? It seems that the likely answer depends upon whether the direction significantly changes the contractual intention of the parties. If a proposed method of working was totally outside the consciousness of the parties then it is likely that this would be considered unreasonable. Such a change in method may well be such that the sub-contractor is no longer competent to carry out the work proposed. Nevertheless, if such a matter was covered by a supervising officer's instruction and issued by the contractor then it is deemed to be a reasonable direction under clause 5.2.2.

In order for the instruction to be deemed a reasonable direction the supervising officer is required to have issued the instruction under the main contract. Does this mean that the supervising officer requires the express power under the conditions of contract, or does it simply mean any instruction issued to the contractor even though the supervising officer has no express power to issue such instructions? This distinction is significant in that if the former is the case the sub-contractor may not be obliged to accept that the instruction be deemed a reasonable direction of the contractor where the supervising officer issues an instruction outside the express powers conferred by the main contract.

Notwithstanding the fact that there is no privity of contract between the employer and the sub-contractor on the terms of the main contract, the sub-contractor would be able to look to the main contract to see if such an instruction is in order. If, on the other hand, the interpretation refers to any instruction then it is deemed a reasonable direction and the sub-contractor is obliged to comply.

The logical and equitable interpretation is that the supervising officer requires express powers under the contract. However, such power can be deemed to be present in the circumstances where an instruction is issued, the contractor requests the supervising officer to specify under what condition such power exists, and following the specification of the clause by the supervising officer the contractor complies notwithstanding the fact that no such express power exists. This in effect leaves the sub-contractor to comply with instructions of the supervising officer which have been issued by the contractor even though the contract has no express provisions to cover the particular instruction. The sub-contractor is therefore left in the hands of the main contractor, and although he can attempt to influence him in this respect the sub-contractor has no legal status in the matter.

Where the supervising officer issues an instruction to the main contractor but has no power under the express provisions to do so, the main contractor may nevertheless comply. (For a discussion of this point see pages 35–8 of the author's book, *Variations in Construction Contracts*.) In these circumstances, does the main contractor's waiver of his own rights bind the sub-contractor to compliance with the direction? It is submitted that they do not and that the instruction not being in accordance with the main contract cannot be deemed a reasonable direction under clause 5.2.2 of NAM/SC.

The contractor may under clause 5.2.1 issue any reasonable direction in regard to the sub-contract works. This clause must be read in conjunction with clause 5.1 and in particular with the words:

'regulating for the time being the due carrying out of the Works ...'

It therefore appears that although the matters upon which directions can be given is very wide it is by no means total. For example, a direction to exclude a person from the site does not fall under the general ambit of this clause.

Variations

There is no express reference (as there is in DOM/1, clause 4.2.1) that a direction includes a variation. Nevertheless, it is clear that 'direction' is intended to include variation and that under clause 5.2.3:

'No variation required by the Contractor or subsequently sanctioned by him shall vitiate this Sub-Contract.'

What constitutes a variation is defined in clause 1.3 and is in words which are similar to those used in JCT 80 or IFC 84. (See chapter 4 of *Variations in Construction Contracts* for a discussion on definition and the extent to which this restricts the supervising officer or contractor in changing his mind.)

Compliance with contractor's directions

The sub-contractor shall comply immediately, so far as that is possible, with all directions which it is within the power of the contractor to give. However, the sub-contractor need not comply with a variation that falls under paragraph 2 of clause 1.3 to the extent that he makes a reasonable objection to such compliance. Frivolous objections would not suffice, and an objection could perhaps only be sustained in circumstances where the variation imposed obligations beyond those reasonably foreseeable and/or substantially affected the way in which the contractor was required to perform the works.

If the sub-contractor fails to comply with any directions which are given by the contractor in accordance with the conditions of contract, save those to which reasonable objection can be sustained, the contractor should give written notice under clause 5.4 of non-compliance. The notice of non-compliance can be given at any stage once a reasonable time has elapsed from the issue of the direction, and if non-compliance continues thereafter for seven days the contractor is entitled to employ others to perform the work.

In accordance with clause 5.4 the contractor is also entitled to recover from the sub-contractor the costs of employing others to perform the work which the sub-contractor had failed to perform and for which he was contractually obligated. This clause is not sound because it does not cater for two distinct situations that may arise:

(1)   where non-compliance relates to an item which the sub-contractor has priced or is deemed to have priced in the sub-contract sum.
(2)   where non-compliance relates to an item which the sub-contractor is obligated to perform but for which he would be entitled to further remuneration in accordance with the sub-contract.

Clause 5.4 adequately provides for the first situation. It is appropriate that where the sub-contractor fails to perform work which he has already priced then the cost of employing others should be deducted. The second situation should only require that the *additional* costs of employing others should be paid, but the clause does not provide for this. Notwithstanding the wording of the clause, it is submitted that this distinction should be made and dealt with as discussed here.

Opening up the work or testing for defective work

Clauses 5.5 and 5.6 of NAM/SC make similar arrangements to those introduced into the Intermediate Form in an attempt to regulate and control the quality of the work. These provisions are very significant and the implementation of the clauses needs to be fully appreciated.

Clause 5.5 enables the contractor to issue directions to the sub-contractor in respect of opening up covered work for inspection or to arrange for or carry out any test of any materials or goods regardless of whether they have been incorporated into the works.

This clause is not well drafted in that it provides for options: either a direction to inspect any work covered up, or to arrange for or carry out any test of materials or goods (whether or not already incorporated in the works).

Strictly speaking this excludes the possibility of uncovering work and then additionally testing that work which has been incorporated. It is not possible to carry out any test of materials or goods which requires firstly that the work be uncovered because this, it seems, would require a specific direction which can only be given as an alternative and not as complimentary to this requirement.

In practice it seems that the contractor would only issue such directions to the sub-contractor on having been instructed himself by the supervising officer, unless the contractor was absolutely certain that the works or goods concerned were defective. This is not a matter of the contractor wishing to wash his hands of the sub-contractor and his responsibilities in this respect. This he cannot do, but he would be very concerned about issuing directions on his own behalf because if no failure was established the cost would fall upon himself.

Many contracts will make provision in the documents for testing of the works. This is usually done by the inclusion of a provisional sum which can be included in either the main contract or the sub-contract documents or indeed in both sets of documents. Where instructions are issued in respect of tests covered by a provisional sum the cost of those tests is set against the provisional sum. Jones and Bergman in their book *A Commentary on the JCT Intermediate Form of Building Contract* at page 159 state:

> 'on a literal interpretation, it appears that the employer pays even if the opening up or testing discloses a failure to meet the contractual requirements ...'

This observation is seductive, but it is submitted that if it were tested in the courts it would be established that if such tests did disclose a failure *on the part* of the contractor then a breach of contract would have arisen; and even though the contractor might be entitled to payment against the provisional sum, the other party would be entitled to damages at least equal to these costs in respect of the breach that had occurred.

Furthermore, the observation itself made by Jones and Bergman is of doubtful validity in that clause 5.5. (clause 3.12 of IFC) provides that:

'The cost of such opening up or testing (together with the cost of making good in consequence thereof) shall be added to the Sub-Contract *unless* provided for in the Priced Documents *or* the inspection or test shows that the materials, goods or works are not in accordance with this Sub-Contract.'

This can be interpreted as Jones and Bergman have done by saying that 'this clause does not apply where the contractor has been required to price for specific testing within the contract sum and which does not stem from an architect's instruction. It appears therefore to relate to the expenditure of a provisional sum ...' and therefore the cost of the instructions issued against this provisional sum will include all costs of testing and making good, even if the opening up or testing discloses a failure to meet the contractual requirements. They do, however, suggest that this may be avoided if careful consideration is given to the wording attached to the description in the provisional sum and clause 1.3 is borne in mind.

Jones and Bergman's analysis can be questioned on a number of counts.

Firstly, the contract may provide for testing and (not by the use of a provisional sum) may still require an instruction. For example:

'The contractor shall allow for providing 6 no. samples of facing bricks and testing for ... as instructed by the supervising officer.'

Secondly, where a provisional sum or some other item of testing (upon which instructions are issued) is included in the contract, clause 3.12 (IFC) and clause 5.5 (NAM/SC) can be interpreted to show that the cost of opening up (although provided for in the contract) is not allowable where the inspection shows that the work etc. is not in accordance with the contract.

In other words the clause when read as a whole gives the following possibilities:

Where an item for testing is provided in the contract documents.

(1)   cost of test and making good to fall on the employer if item priced by the contractor and where test shows that the contractor has complied with the contract.

(2)   cost of making good to fall on the contractor where test shows that the contractor has not complied with the contract (the cost of the test itself will probably fall on the employer as an additional item unless a provisional sum is used but it should more properly fall on the contractor).

Where no item for testing is provided in the contract documents.

(1)   cost to fall on the employer where test shows that the contractor has complied with the contract.

(2)   cost to fall on the contractor where test shows that the contractor has not complied with the contract.

Simply because an item for testing is included in the contract this does not mean that one cannot look to the alternative proviso, i.e. 'or the inspection or test shows that the materials, goods or works are not in accordance with the sub-contract'.

The author is certain that this is the intention and submits that this is the correct interpretation of the clause but does concede that the drafting of the clause is not wholly sound. The suggestion that the wording of the provisional sum requires careful consideration is useful and of greater impact when incorporated in a named sub-contract because of the absence of clause 1.3 (IFC 84).

The main contractor would be advised to provide adequate supervision to ensure that the sub-contractor provides work in accordance with the contract, and if he felt that testing etc. might be required at his instigation then this should be provided for in the sub-contract documents. It is perhaps better that the supervising officer prescribes such provisions in the sub-contract inasmuch as it puts the tender submission on the same basis without the contractor being asked to take what may be considered unreasonable risks. Where the supervising officer does make such a provision it may be wise to extend its operation to cover materials and goods brought on to site by the sub-contractor where the ownership is not vested in the sub-contractor.

The next problem is with clause 5.6 and concerns the determination of what is a failure. This issue is of great significance because if the work fails the cost of test and the consequential making good will either directly or indirectly fall upon the sub-contractor. Generally, one considers that the determination of failure is a straightforward matter but this is not the case because:

(1)   there may be tolerances (which may or may not be established) in which the work falls.
(2)   the test may prescribe standards beyond those for which the sub-contractor has contracted.
(3)   the failure may be accepted but its extent is unknown.

Therefore, in deciding whether a test or inspection establishes failure on the part of the sub-contractor one should consider these issues.

The consequences of a failure of the sub-contractor's work is not restricted to making good the work which has been shown to have failed. The consequences extend much further because of the introduction of requirements to show that other work has not failed. Clause 5.6.1 requires that the sub-contractor upon the discovery of failed work shall set out what action he proposes to take (at no cost to the contractor) to establish that there is no similar failure.

The sub-contractor must inform the contractor of his proposals within five days of discovery of the initial failure. If the sub-contractor fails to do this or his proposals are not satisfactory to the contractor or because of safety or statutory obligations the contractor is unable to wait the five days, then the contractor may himself issue directions as to what should be done.

The major difficulty in this situation is establishing what is meant by 'no similar failure in work'. Is this really intended to mean what it says? Such words could embrace the following:

(1)   where concrete is found to have been mixed in the wrong proportions all other work involving mixing could fall to be tested.
(2)   where a technical failure (i.e. not complying with the specification) is established all other work may be tested to see if this technical failure exists elsewhere.

It is doubtful whether this is the intention, but if this was not intended it would have been perhaps more appropriate to use the words 'no failure in similar work'. In practice it is likely that the clause will be operated as though these words existed in the contract, but clearly they do not and as a consequence there are going to be occasions when the proposals put forward by the sub-contractor are not going to be acceptable to the contractor.

As already noted, where this occurs the contractor may issue directions as to what must be done and the work involved must be carried out forthwith and at no cost to the contractor. The sub-contractor may well be put to great expense as a consequence of these directions, and although he is obliged to comply with the directions he can within seven days of their receipt write to the contractor stating that he objects and giving his reasons why. The contractor, upon consideration of the sub-contractor's submission, may withdraw or modify the instruction and thus overcome the sub-contractor's objection.

If this is not done within seven days of the sub-contractor's notice then the matter shall be referred to an arbitrator in order to establish what is reasonable in all the circumstances. Interestingly, although the issue referred to the arbitrator is stated to be 'what is reasonable in all the circumstances' clause 5.6.5 refers to the arbitrator deciding what is 'not fair and reasonable', and where he so decides the sub-contractor is entitled to payment and possibly an extension of time.

From this it can be seen that there is an attempt to protect the sub-contractor from his own initial shortcomings. However, the protection afforded is not particularly reliable in the light of the possible interpretations of 'no similar failure in work'. How and what the arbitrator will decide is fair and reasonable is extremely debatable. One thing which is certain is that he should not be influenced by the result of the further tests, although where all the subsequent tests have indicated a failure the sub-contractor will either be caught under these provisions or sued for damages for breach of contract.

A further complication can arise where the contractor decides to withdraw or modify his directions following an objection by the sub-contractor but where the sub-contractor has already carried out the inspection because he is obliged under clause 5.6.3 to comply forthwith. It would appear that the matter would not then fall to be a dispute as referred to under clause 5.6.4 but would nevertheless be referable to arbitration under clause 35.

Taking into account all the difficulties that may arise as a consequence of a failure of work, it is prudent for the sub-contractor to spend more time ensuring that a failure does not occur rather than attempting to escape the effects of these onerous clauses.

Removal of defective work

IFC 84 provides in clause 3.14 that the supervising officer may issue instructions with regard to the removal from site of any work, materials or goods which are not in accordance with the contract. This is the same clause that is contained in JCT 80 (clause 8.4) and JCT 63 (clause 6(4)) and that received judicial comment in *Holland, Hannen & Cubitts (Northern) Ltd* v. *Welsh Health Technical Services Organisation and Others* (1981). In common with other sub-contract forms, no such clause is specifically included in NAM/SC. Nevertheless, the sub-contractor could be required to remove defective work either by the contractor himself under clause 5.2.1 or by the forwarding of an instruction of the supervising officer under clause 5.2.2 which has been issued in accordance with clause 3.14 of the main contract.

Access to the works

Building contracts often expressly provide that the supervising officer shall have a right of access to the works. This is not strictly necessary because, without doubt, the right of the supervising officer to have access to the works would be implied. Furthermore, although it may be in the mind of the contractor to want to exclude the supervising officer this would generally be somewhat foolhardy. The supervising officer and contractor should be working towards the common objective of providing the building. All of this was no doubt in the mind of the Joint Contracts Tribunal when they decided to omit such a clause from the Intermediate Form of Contract.

If such a right can be implied then this will enable the supervising officer to obtain access to the works for the purposes of inspecting the sub-contract work, because the main contract works incorporate the sub-contract works. However, such an implied right in the main contract would not extend to the supervising officer having access to the workshops or other places of the contractor where work is being prepared. Therefore, one can appreciate the need for words such as are included in JCT 80, clause 11:

'The architect/supervising officer and his representatives shall at all reasonable times have access to the Works *and to the workshops or other places of the Contractor where work is being prepared for this Contract ...*'

These words do not extend to cover the workshops etc. of sub-contractors and therefore in an attempt to provide access to such places the clause goes on to state:

'... the contractor shall by a term in the sub-contract so far as is possible secure a similar right of access ... and shall do all things necessary to make such a right effective.'

John Parris, in *The Standard Form of Building Contract: JCT 80*, states that he believes this is wide enough to require the contractor to institute proceedings against a sub-contractor if the sub-contractor refuses admission to the premises. Clearly, the supervising officer has no direct right against the sub-contractor because no contract exists between them.

The position under the Intermediate Form of Contract and NAM/SC is somewhat different. Firstly, no clause similar to clause 11 exists and therefore the only access that can be implied is access to the works themselves.

Clause 23 of NAM/SC attempts to give a wider right of access in respect of the sub-contract works. The clause reads:

'The Contractor and the Architect/the Supervising Officer and all persons duly authorised by either of them shall at all reasonable times have access to any work which is being prepared for or will be utilized in the Sub-Contract Works.'

This is strange because it seems necessary to provide such a right in respect of the sub-contract works but not in respect of the main contract works.

However, notwithstanding this clause the supervising officer cannot insist on admission to the sub-contractor's premises because, as previously noted, there is no privity of contract under which this right can be enforced. The contractor has such a right under this clause because NAM/SC is a contract between the parties concerned, but there is no way the supervising officer can insist upon admission. In the absence of words such as '... shall do all things necessary to make such a right effective' in the main contract there is no lever which can be used against the contractor to enforce his rights against the sub-contractor.

A sub-contractor would generally accommodate the supervising officer with regard to allowing him access, but the supervising officer should be aware that the sub-contractor is not obliged to do so and furthermore that it generally costs the sub-contractor to provide such facilities.

It should also be noted that the contractor can authorise other persons to have access to any work but once again the supervising officer cannot insist that such authority is given under the contract. In practice, however, if the sub-contractor proved intransigent with regard to granting such access to the supervising officer it seems likely that the contractor would give the supervising officer authority under clause 23.

Assignment and sub-letting

Clause 24.1 states:

'The Sub-Contractor shall not without the written consent of the Contractor assign the sub-contract.'

The exact meaning of this type of clause, which is standard to many building contracts, has been questioned by a number of eminent lawyers. For example, both Donald Keating in *Building Contracts* and John Parris in *The Standard Form of Building Contract: JCT 80* observe that the word 'assign' is probably not used in its strict sense.

It is not intended to consider this issue at length here, but it is worthwhile expressing what may be deduced to be the position in law.

(1)   A sub-contractor cannot assign the burden of a contract, with or without the consent of the other party; however, it can be agreed that the sub-contractor ceases to be liable for his performance and that someone else takes his place. This is known as novation and is not to be confused with assignment.

(2)   A sub-contractor may assign the benefits of a contract by way of a legal or equitable assignment without necessarily obtaining the consent of the contractor.

(3)   A prohibition to assignment of the benefits of a contract is generally therefore not effective; however, such a clause may be supported by a clause which provides for determination if the assignment clause is not complied with. This would then give a remedy to the contractor for a breach, but it is submitted that it does not necessarily affect the assignment which has taken place.

(4)   'Assign' in this clause probably is intended to mean vicarious performance. Vicarious performance is the execution by another,

i.e. a sub-sub-contractor, of works which have been contracted for by the sub-contractor. The sub-contractor, however, remains liable for the whole of the sub-contract works.

(5) Vicarious performance may be in respect of the whole or part of the work, and it may be possible in certain situations to secure vicarious performance without the consent of the other party; but not under NAM/SC, as this contains an express clause which requires such consent to be obtained.

As vicarious performance may be in respect of the whole or part of the work one is left wondering what is the purpose of clause 24.2 which refers to sub-letting.

From the foregoing it is possible to conclude that vicarious performance of the whole works is intended to fall under clause 24.1. This is on the basis that the clause refers to 'assign the Sub-Contract' and that the word 'assign' has been used as described in item (4) above. Consequently, if such consent is withheld no reason need be given. It is simply a matter of whether the contractor chooses to give his consent.

Vicarious performance of part of the work falls under clause 24.2. The sub-contractor may choose to sub-let any portion of the sub-contract works so long as he secures the consent of the contractor. However, it should be noted that such consent shall not be unreasonably withheld. This proviso could cause difficulties in practice because it seems to imply that unless the contractor has a good reason for refusing consent the sub-contractor is able to sub-let the works. If this is so then the common law position is somewhat modified in that the contractor would have to establish that vicarious performance should not take place because the contract is personal in nature.

A sub-contractor has no rights to sub-let works where there is a contract clause prohibiting sub-letting or where the contract is personal in nature. This has been established in the cases of *British Waggon Co.* v. *Lea* (1880), *Davies* v. *Collins* (1945) and *Edwards* v. *Newland and Co.* (1950). Therefore, one can see that clause 24.2 on the one hand prevents sub-letting by prohibiting it without consent, and on the other permits sub-letting where consent is given and such consent must not be withheld unreasonably.

A named sub-contractor comes into existence because the designer wishes to secure a particular sub-contractor to execute a specific part of the works. Therefore, logically there should be good reason for naming a sub-contractor, and if such a good reason exists the chances are that it supports the view that the contract is personal in nature.

It seems, therefore, that the only good reason for refusing consent is that the contract requires the sub-contractor's skill or special knowledge. This is of course very wide in interpretation and it may well include management expertise as well as technical expertise. However, it seems that this reason should always exist, otherwise why have a named sub-contractor? Therefore, the clause should surely prohibit sub-letting *per se*. As it would be logical to do so in these circumstances it may be construed that some other reason needs to be established in order for consent to be withheld. What this would be it is difficult to imagine.

It is interesting to note that clause 24.2 requires the consent of the contractor, and presumably he would argue the case of the supervising officer if consent was to be withheld. Whether the contractor can give his consent without reference to the supervising officer is, it seems, debatable.

Although the contractor cannot without consent sub-contract any part of the works, sub-contracting in accordance with clause 3.3 (named persons as sub-contractors) is excluded. Whether such exclusion extends to any sub-letting by the named sub-contractor is unclear, but surely it is not intended to embrace this form of sub-letting. The contractor would be unwise to give consent to a sub-contractor without the authority of the supervising officer, but once the contractor has given his consent the sub-contractor can rely upon it regardless of whether the contractor sought authority or not. In spite of the confusion created by clause 24, a prohibition to assignment must be distinguished from a prohibition to sub-contracting because they are distinct issues in law.

Unfortunately, the case of *Thomas Feather and Co. (Bradford) Ltd* v. *Keighley Corporation* (1953) only adds further confusion to the situation where a breach of a prohibition of sub-contracting has occurred. This case decided *inter alia* that such a breach of a condition would not have entitled the corporation to determine but for the existence of a separate express provision permitting such determination; furthermore, that where the contract was so determined no damages could be recovered. This raises a number of important issues, for example:

(1)   does this mean that the sub-contractor can invite the contractor to determine the sub-contract without suffering damages? (A sub-contractor may well wish to do this when performing an uneconomic contract.)

(2)   does this mean that a sub-contractor can almost with impunity ignore a clause prohibiting sub-letting where it is not supported by a determination provision?

It is submitted that the decision in the *Thomas Feather* case should not be construed too widely because the contract gave an option to determine or require liquidated and ascertained damages in respect of the breach of the sub-letting clause. It is this option which surely distinguishes this contract from other contracts. For further criticism of this case see John Parris's book *Default by Sub-Contractors and Suppliers* at pages 4–5.

In conclusion one would advise the sub-contractor always to seek consent before sub-letting part of the works. If consent is withheld and the sub-contractor feels strongly enough about it to pursue the issue, there is at least an arguable case.

Chapter 7

General obligations and insurance – NAM/SC

Liability under main contract

Clause 6.1.1 is a wide ranging clause which attempts to impose upon the sub-contractor similar liabilities to those which the main contractor accepts under the terms of the main contract, in so far as they apply to the sub-contract:

> 'The Sub-Contractor shall:
>   observe, perform and comply with all the provisions of the Main Contract as referred to in NAM/T, Section 1, on the part of the Contractor to be observed, performed and complied with so far as they relate and apply to the Sub-Contract (or any portion of the same ...)'

This clause is an attempt to reduce the volume of the sub-contract by avoiding the repetition of clauses. Unfortunately, as with most short cuts a price has to be paid. In this case one is unsure as to the precise extent of the sub-contractor's obligations. The intention is reasonably clear in that the sub-contractor is presumed to take on the like liabilities of the contractor inasmuch as they are relevant. Therefore, from a practical point of view the sub-contractor is best advised to accept this as the position.

The wording used in this clause is not as sound as it might be in that it says '... all the provisions of the Main Contract as referred to in NAM/T, Section 1 ...' whereas it perhaps would be better to say '... with all the provisions of the Main Contract as described by or referred to in the NAM/T, Section 1'. The latter wording is similar to that used in clause 5.1.1 of NSC/4 and avoids the possibility of arguing that the only provisions of the main contract which the sub-contractor must account for are those which are expressly identified in NAM/T, section 1. Even if this defect in wording is material its significance is greatly reduced by the fact that the clause goes on to identify, without prejudice to the generality of clause, the specific clauses which the sub-contractor must observe, and furthermore clause 6.1.2 requires a wide ranging indemnity to be given by the sub-contractor.

Clause 6.1.1 therefore identifies the major clauses in the Intermediate Form of Contract which are relevant to the sub-contract as:

Clause 1.10   Passing of property in unfixed materials or goods.
Clause 1.11   Passing of property in off-site materials.
Clause 3.9   Levels and setting out.
Clause 5.1   Statutory obligations, notices, fees and charges.
Clause 5.2   Notice of divergence between statutory obligations and contract documents.
Clause 5.3   Extent of contractor's liability to employer for non-compliance with statutory obligations.
Clause 5.4   Compliance with the needs of an emergency.

One wonders why a clause starts off by including all the provisions of the main contract and then specifically mentions the major clauses. This gives the impression that one has gone as far as one can be bothered to go in identifying the relevant clauses, and just to make certain one then covers all the other clauses which may or may not be relevant. This generally does not alter the intended legal position, so the only justification for it is in order to communicate more clearly. However, this is not a good means of communication and it is arguable that the major issues need not be referred to as they will be well known whereas it is the lesser known issues that need bringing to the attention of the sub-contractor if the purpose of such reference is communication.

The items which may be relevant and not expressly referred to include, for instance:

Clause 1.1   Contractor's obligations.
Clause 2.10   Defects liability.
Clause 3.4   Contractor's person-in-charge.
Clause 3.11   Work not forming part of the contract.
Clause 3.14   Instructions as to removal of work not in accordance with the contract.
Clause 6.2.1 Sub-contractor's limit of insurance cover.
Clause 6.2.3 Sub-contractor not insuring.

It can be seen from this that very significant issues are included, such as what might be the sub-contractor's obligations and what is the defects liability, and less important matters, such as should the sub-contractor keep a person-in-charge on site and the employer's right to insure where the sub-contractor has failed to do so. It is therefore extremely difficult to appreciate the rationale for certain clauses being referred to specifically in clause 6.1.1.

Passing of property in goods

The sub-contractor's obligations in respect of the passing of title in goods are dealt with on pages 168–170 and page 204.

Levels and setting out

Under clause 3.9 of IFC 84 the contractor is made responsible for correcting errors arising from his own inaccurate setting out at no cost to the employer. This need hardly be expressed because the contractor would in any event be responsible for providing work in accordance with the contract documents. Nevertheless, the sub-contractor is also made aware of it under clause 6.1.1. The clause does, however, go further by allowing the supervising officer, with the consent of the employer, to instruct that the error shall not be amended and that an appropriate deduction for such errors not required to be amended shall be made from the contract sum. This clause is clearly intended to apply to sub-contractors as well and gives the supervising officer/employer an option not otherwise available to him in law.

The wording of clause 3.9 is an improvement on the similar clause 7 in JCT 80 and removes any possibility that this clause could be interpreted to mean that the contractor could be paid for the correction of his own errors by the words 'unless ... otherwise instructed ...'.

However, where the sub-contractor is instructed by the supervising officer through the contractor that errors arising from inaccurate setting out should not be remedied, an *appropriate deduction* shall be made. A decision as to what is an appropriate deduction is not that easy.

It can be argued that the deduction equals the cost that would be incurred if the error was remedied. This in broad terms corresponds to how the error may be perceived if damages were being sought for breach of contract. But it is difficult to reconcile this with the situation where the error has given an advantage to the employer, and in these circumstances it is suggested that no appropriate deduction should be made. Where the cost of remedying the error is less than the damage suffered by the employer, the error should be corrected because it is unlikely that the employer will be able to secure the greater amount where the supervising officer instructs that the error should not be remedied.

Statutory obligations

The sub-contractor is put under the same obligations as the contractor in respect of complying with all statutory requirements and the giving of notices, giving notice of discrepancies between the statutory requirements and contract documents, and in respect of emergency work necessary to satisfy statutory regulations.

Although clause 5.1 (IFC 84) also requires the contractor to pay all fees and charges in respect of statutory requirements, the contractor is able to recover the cost of these fees from the employer by an adjustment of the contract sum unless they are already provided for in the contract documents.

If the sub-contractor carries out work in accordance with the contract documents but that work does not comply with the statutory requirements then the sub-contractor, by virtue of clause 5.3 (IFC 84), is relieved of any liability to the employer in respect of the non-compliance. This clause is subject to clause 5.2 (IFC 84) which requires any divergence which the contractor finds to be notified to the supervising officer. Obviously, difficulties arise in deciding whether the contractor and/or sub-contractor did or did not find the divergence, but superficially it seems that no notice means no divergence has been found.

However, there comes a point when the sub-contractor could not reply upon this for there is little doubt that there is an argument that the sub-contractor must surely have known that he was executing work not in accordance with the regulations. Notwithstanding this, the sub-contractor is not relieved of his statutory obligations *per se* and although he may not be liable to the employer he may still be liable under the Act of Parliament.

For example, in *Street and Another* v. *Sibbabridge Ltd and Another* (1980) it was established that the building regulations overrode the express conditions of contract and that it is the contractor who commits a criminal offence if the regulations are not complied with. This statement of the law puts an onerous obligation upon the contractor and therefore on the sub-contractor as well. It is apparent that one or the other would be responsible for remedying work not in accordance with the building regulations but that the cost of the remedial work might fall upon the employer if the sub-contractor had executed the work in reliance upon the contract documents.

Where an emergency arises in fulfilling the statutory obligations under clause 5.1 (IFC 84) the sub-contractor is required to perform the emergency work and to inform the contractor immediately.

Exclusions from liability

Although clause 6 imposes a variety of obligations upon the sub-contractor it also, in clause 6.2, makes clear the exclusions from such liability:

'Nothing contained in the Sub-Contract Documents shall be construed so as to impose any liability on the Sub-Contractor in respect of any act, omission or default on the part of the Employer, the Contractor his other sub-contractors or their respective servants or agents ...'

This clause purports to cover matters much wider than negligence for it states that the sub-contractor is not to be liable 'in respect of any act, omission or default'. There is some doubt as to the meaning of these words, as there is with the words 'act or neglect' (this term is discussed at page 104. However, the words 'act or neglect' have been given judicial consideration whereas no such consideration has been given to 'act, omission or default'. It does seem that in the context of this clause the latter would be taken to include any act, omission or default regardless of whether they are negligent. An omission or default does not in itself constitute negligence. Therefore, if this submission is correct the sub-contractor does not adopt or accept liability under the terms of the contract for work done by others or for the omission or default of others, even though such omission or default is not negligent.

It does not follow, however, that the sub-contractor is free of all liability in respect of these acts. There is still the possibility of claims in tort and under certain statutory provisions. It is not within the ambit of this book to discuss these issues at length, but it is necessary to indicate that such a clause does not absolutely restrict liability.

Actions against the sub-contractor can still arise in a variety of ways, for example:

(1) by a third party against the sub-contractor as a joint tort feasor.
(2) by the employer, other sub-contractors or their respective servants or agents in tort against the sub-contractor because the sub-contractor owed the other a duty to warn about the acts he had performed (the contractor is unlikely to succeed because of the existence of a sub-contract and a clause directly related to this issue).
(3) a statutory obligation.

In law liability for the negligence of others can occur by way of vicarious liability or under an indemnity given under a contract to that person in respect of his own negligence.

It is submitted that although clause 6.2 is fairly restrictive it is nevertheless absolute protection against claims by the contractor under the contract in respect of these matters. Lord Justice Devlin said in *Walters* v. *Whessoe Ltd and Shell Refining Co. Ltd* (1960):

'It is now well established that if a person obtains an indemnity against the consequences of certain acts, the indemnity is not to be construed so as to include the consequences of his own negligence unless those consequences are covered either expressly or by necessary implication. They are covered by necessary implication if there is no other subject matter upon which the indemnity could operate.'

The indemnity given does not expressly provide to cover the contractor's own negligence. In fact the proviso in clause 6.2 ensures that the indemnity specifically excludes such consequences.

Clause 6.2 also goes on to state:

'... nor create any privity of contract between the sub-contractor and the employer or any other sub-contractor.'

The validity of this clause is dubious. It purports to ensure that the sub-contract documents do not create privity of contract between the sub-contractor and the employer or other sub-contractors. This it cannot do because the employer and other sub-contractors are not bound by this sub-contract and therefore are not bound to accept that privity does not exist. In other words, although it may be the intention not to create privity of contract between the sub-contractor and the employer or other sub-contractors upon anything contained in the sub-contract documents, it is clearly possible that such privity can be created and this clause does nothing to prevent it occurring. ESA/1 is a clear indication that a contract can exist albeit in different terms.

Perhaps the real purpose of the clause is to emphasise that the employer and other sub-contractors are not parties to this sub-contract. If this is what is intended then perhaps it should say so.

Indemnity to contractor

The existing indemnity and insurance provisions of the JCT Standard Forms of Contract will all be substantially affected by the JCT Amendments issued in November 1986. As it will be some time before these revisions apply to all contracts it is intended firstly to deal with the contracts as existing and then separately to deal with the effect of the JCT Amendment 1 published by BEC Publications in respect of NAM/SC. A separate Amendment 1 in respect of IFC has been published by the RIBA.

In addition to the obligations imposed by clause 6.1.1 the sub-contractor is required by clause 6.1.2 to indemnify the contractor against and from:

'any breach, non-observance or non-performance by the sub-contractor or his servants or agents of any of the provisions of the main contract insofar as they relate and apply to the sub-contract and any act or omission of the sub-contractor or his servants or agents which involves the contractor in any liability to the employer under the provisions of the main contract insofar as they relate and apply to the sub-contract.'

Dennis Turner in his book *Building Sub-Contract Forms* states in the context of the NSC/4 sub-contract that all of these matters arise out of the liabilities of the sub-contractor under the main clause (5.1.1 of NSC/(4). This is not strictly true. Although clause 5.1.1 (WSC/4) identifies the liabilities imposed by the obligations contained in clause 5.1.2, without the existence of the indemnity clause the duties would only be owed to the contractor in respect of the breach under the sub-contract. Clause 5.1.2 provides a further obligation by providing for an indemnity clause wide enough to cover any liability of the contractor to the employer. Clearly, specific liabilities arise under this clause. It should also be noted that the words used in 6.1.2 (NAM/SC) are preferable to those used in the NSC/4 sub-contract as they leave no doubt that the indemnity extends to liability arising from all the provisions of the main contract. Clause 6.1.2 also goes further because its effect is to require the sub-contractor to indemnify the contractor against any judgment in favour of a third party and furthermore clause 6.1.2.2 covers possible damage and loss which would not necessarily be recoverable by the contractor under the sub-contract without the existence of the indemnity.

The indemnity given under 6.1.2 is much more onerous than the contractual obligation under clause 6.1.1 because, as was reaffirmed in *County and District Properties* v. *C. Jenner and Son Ltd* (1976), the cause of action under the indemnity does not arise until a loss has been established. This in effect increases the limitation period and a contractor may be caught by an indemnity clause where he might otherwise have escaped.

Clause 6.1.2.1 does, however, overlap substantially with clause 7 which deals with injury to persons and property, but in the light of *City of Manchester* v. *Fram Gerrard Ltd* (1974) it appears that clauses 6.1.2.1 and 6.1.2.2 would not cover sub-sub-contractors whereas clause 7 would.

Clause 6.1.2.3 is a strange clause, and the similar clause in DOM/1 and the liability it creates is referred to in *Building Sub-Contracts Forms* in the following terms:

'This arises quite apart from any liability occasioned by the main contract conditions and it is difficult to see that the matter is not covered by the generality of domestic clause 6 about indemnity over injury to persons and property.'

(The clause 6 referred to is in DOM/1 but the clause is the same as clause 7 contained in NAM/SC.)

This statement certainly appears to represent the case, and because clauses 6.1.2.3 and 7 overlap one may be left with a small problem should a matter of negligence on the part of the sub-contractor arise. The problem is, does the contractor have to establish negligence to secure his indemnity under clause 6.1.2.3, or does he simply have to show that he has suffered loss and that it was not caused by himself or the others referred to in clause 7 and claim his indemnity under this clause?

The indemnity given by the sub-contractor under clause 7.1 excludes any act or neglect of:

- the contractor or his servants or agents or of any other sub-contractor
- the servants or agents of other sub-contractors
- the employer or of any person for whom the employer is responsible.

The words 'act or neglect' have already been referred to at page 100. These words may appear to go beyond negligence but there is authority in *Hosking* v. *De Havilland Ltd* (1949) and *Murfin* v. *United Steel Companies* (1957) that these words do not apply to a breach of statutory duty which does not amount to common law negligence. The indemnity given does not then appear to go beyond negligence and therefore it is only those acts or neglects which are negligent that are excluded from the indemnity given by the sub-contractor to the contractor.

It should be noted that clause 7.1 puts the onus of proof on the sub-contractor. In other words, the sub-contractor must show he is not liable because it is an act or neglect of the contractor etc.

Under clause 7.2 the sub-contractor indemnifies the contractor against claims in respect of injury to real or personal property which are caused by the sub-contractor's negligence, omission or default. This indemnity includes the sub-contractor's servants and agents, but under this clause the onus of proof shifts to the contractor and requires him to establish that the claims are on account of the sub-contractor's negligence, omission or default.

Clauses of this nature are sometimes referred to as giving indemnity for the sub-contractor's negligence or similar act. This could be misleading because the similar act suggests that although not negligent it is close to being negligent whereas the clause makes the sub-contractor responsible for acts of omission or default regardless of negligence as well as for the sub-contractor's negligence.

The use of the words negligence, omission or default has been criticised by Frank Eaglestone in his book *Insurance Under the JCT Forms* where it is given full consideration at pages 37–39.

The major exceptions to the indemnity given under clause 7.2 is in respect of loss or damage which arises under clause 9.1.1.1 or which is at the sole risk of the employer under the relevant main contract provisions. This exception has the effect of relieving the sub-contractor of any liability to the contractor in respect of the IFC 84 'Clause 6.3 Perils' which are defined in clause 8.3 (IFC 84). Therefore, the sub-contractor is not responsible under the sub-contract for loss or damage occasioned by these perils even if caused by the sub-contractor's own negligence.

Responsibility for sub-contractor's plant

The sub-contractor under clause 11.1 is understandably made solely liable for any loss or damage to or caused by:

> 'The plant, tools, equipment or other property belonging to or provided by the sub-contractor, his servants or agents and any materials or goods of the sub-contractor which are not properly on site for incorporation in the sub-contract works ...'

However, the clause does not include loss or damage due to any negligence, omission or default of the contractor, his servants or agents or sub-contractors (other than the sub-contractor himself and his servants and agents).

The words 'not properly on site for incorporation' are a peculiar turn of phrase. They do not mean that plant on site and which should not be there is embraced by these words. They refer to that which is not for incorporation as contrasted with that which is to be incorporated. The latter, under clause 9.3, is still at the risk of the sub-contractor. Plant etc. brought on to site by the sub-contractor but which should not be there is not covered by clause 11.1 but the sub-contractor is still responsible for his actions which are outside the terms of the contract.

Additionally, the sub-contractor shall indemnify the contractor against any expense, liability loss claim or proceedings in respect of loss or damage caused by the plant and tool etc. belonging to themselves or their servants or agents. It would seem that the indemnity does not extend to cover the contractor's own negligence or the negligence of those for whom the contractor is made responsible.

Insurance

The sub-contractor under clauses 6 and 7 gives certain indemnities to the main contractor and it is prudent that he should insure, as far as it is possible, in respect of the possible liabilities under these indemnities regardless of any contractual obligations to insure.

Clause 8 requires the sub-contractor to insure against liabilities arising:

> '... in respect of personal injury or death arising out of or in the course of or caused by the carrying out of the Sub-Contract Works, not due

to any act or neglect of . . . the Contractor or . . . in respect of injury or damage to property real or personal arising out of or in the course of or by reason of the carrying out of the Sub-Contract Works and caused by any negligence, omission or default of the Sub-Contractor, his servants or agents.'

This insurance is without prejudice to the sub-contractor's liability to indemnify the contractor and provides some measure of financial support to the indemnity given. If the insurance was inadequate or failed then the sub-contractor could still be pursued by the injured party. Therefore, the insurance gives the contractor some security should he need to claim against the sub-contractor, but the sub-contractor cannot hide behind the existence of this insurance because he has given an indemnity in clauses 6 and 7 and is responsible for any shortfall.

The insurance cover required by clause 8.1 is complementary to the indemnities given in clause 6 and 7 but does not precisely match the indemnities given and this is shown in fig. 7.1.

Amount of insurance cover

The insurance cover that is required to cover claims for personal injury or death of any person under a contract of service or apprenticeship with the sub-contractor and whilst working for him must comply with the Employer's Liability (Compulsory Insurance) Act 1969 and any statutory orders made thereunder or any amendment or re-enactment thereof. One such set of regulations is the Employer's Liability (Compulsory Insurance) General Regulations 1971 which laid down the extent of restrictions which could be included in employers' liability policies and required that the minimum limit of insurance was £2 million in respect of any one occurrence for which the employer is held responsible. In practice this limit is not generally stipulated as employers' liability policies are issued without a limit being imposed.

Clause 8.2 states that in respect of all other claims which arise under the insurance required by clause 8.1 the sum shall be as stated in NAM/T, section II, item 2. The insurance cover inserted by sub-contractor for any one occurrence or series of occurrences arrising out of one event shall be not less than the amount inserted in the main contract Appendix – clause 6.2.1, which is reproduced in NAM/T, section I, item 6.

Two points emerge from this. Firstly, the reference in item 2 of section II to NAM/SC 7 is erroneous, because this is the indemnity clause and

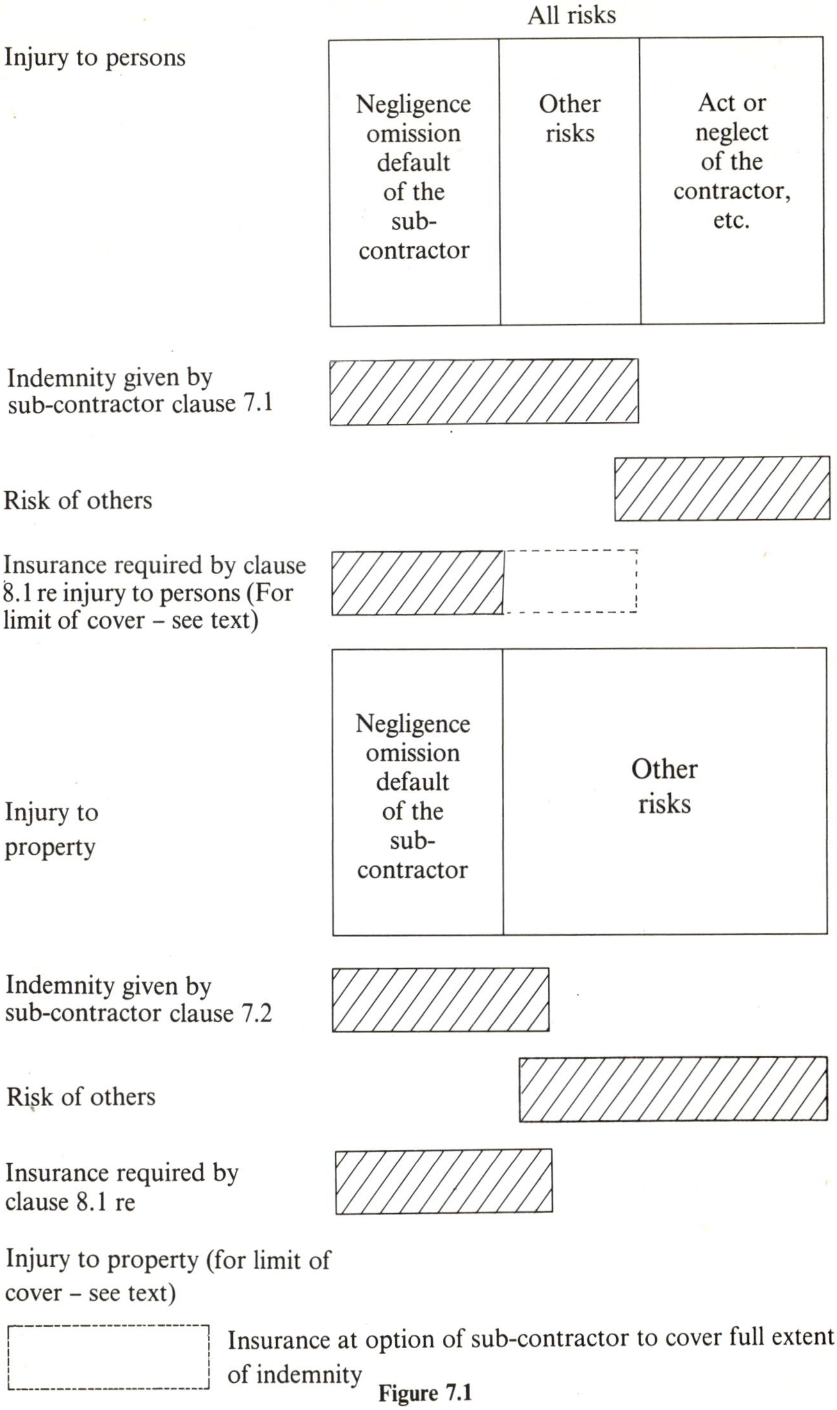

All risks

Injury to persons

| Negligence omission default of the sub-contractor | Other risks | Act or neglect of the contractor, etc. |

Indemnity given by
sub-contractor clause 7.1

Risk of others

Insurance required by clause
8.1 re injury to persons (For
limit of cover – see text)

| Negligence omission default of the sub-contractor | Other risks |

Injury to
property

Indemnity given by
sub-contractor clause 7.2

Risk of others

Insurance required by
clause 8.1 re

Injury to property (for limit of
cover – see text)

Insurance at option of sub-contractor to cover full extent
of indemnity

Figure 7.1

the reference should be to clause 8; and secondly, the sub-contractor may insert in item 2 of section II an insurance cover in excess of the amount stated in the main contract appendix.

One may wonder why a sub-contractor should wish to insert a figure for insurance cover in excess of that contained in the main contract. The reason is fairly straightforward in that the indemnity given is without limit and therefore the sub-contractor may feel that it is commercially advisable to carry greater insurance cover than is required by the main contract.

There is no question that the indemnity is limited solely to the extent of insurance cover provided and therefore the sub-contractor has to make a commercial judgment in respect of the extent of cover (beyond the minimum required) that is affordable. Additionally, he must decide the duration of such insurance because no time period is mentioned in the contract. In practice this will seldom present any problem because the insurance taken out by the sub-contractor will be for all the work performed and insured on a continuing basis as compared with taking out separate policies for each project.

Loss or damage

Clause 9.1 deals with who insures the works and materials and goods on site in respect of loss and damage by the clause 6.3 perils.

The clause 6.3 perils are defined as:

> 'fire, lightning, explosion, storm, tempest, flood, bursting or over-flowing of water tanks, apparatus or pipes, earthquake, aircraft and other aerial devices or articles dropped therefrom, riot and civil commotion. EXCLUDING any loss or damage caused by ionising radiations or contamination by radioactivity from any nuclear fuel or from any nuclear waste from the combustion of nuclear fuel, radioactive toxic explosive or other hazardous properties of any explosive nuclear assembly or nuclear component thereof, pressure waves caused by aircraft or other aeriel devices travelling at sonic or supersonic speeds.'

Under the main contract (IFC 84) there are three options for insuring the work:

- Clause 6.3A   Contractor to insure – new buildings.

- Clause 6.3B   Employer to insure – new buildings.
- Clause 6.3C   Employer to insure – existing buildings.

The sub-contractor is not responsible for any loss or damage caused and to which clauses 6.3A, 6.3B or 6.3C (whichever is applicable) apply. Clauses 9.1.1.2 and 9.1.2.2 relieve the sub-contractor of 'any loss or damage however caused', but this must be read in the context of the clause and only relates to the 'perils' required to be insured under the respective clauses. In other words, 'however the *peril* arises', which may nevertheless include the sub-contractor's own negligence, the sub-contractor is protected under these clauses. The words 'required to be insured' are relevant because it may not be possible to insure against all the perils specified.

One could still, however, argue that if a 'peril' is excluded under the insurance clauses 6.3A, 6.3B and 6.3C then the extent to which the sub-contract is relieved is to the extent of the insurance taken out under these clauses.

Under clause 9.4 the sub-contractor:

'... shall observe and comply with the conditions contained in the policy of insurance of the contractor or the employer, as the case may be, against loss or damage by any of the "clause 6.3 perils".'

What is the impact of this clause upon clauses 9.1.1.2 and 9.1.2.2 if the sub-contractor fails to observe any of the conditions and thus enables the insurer to deny liability under the policy?

It appears that the sub-contractor puts himself at great risk and would be advised to obtain copies of the relevant policies in order to ensure that he observes and complies with its requirements. Therefore, the purpose of clause 9.1 is (1) to provide information in order that the sub-contractor can decide what matters not included in these clauses need to be insured against and (2) to enable clause 9.2 to operate in respect of claims for loss and damage without further reference to other documents.

If the main contractor is responsible for insuring a new building clause 9.1.1.1 will apply and the insurance will cover the clause 6.3 (IFC 84) perils.

The insurance will cover loss and damage caused by these perils to all work executed and all unfixed materials and goods which have been delivered to, placed on or are adjacent to the works and which are intended to be incorporated into the works. The insurance cover required by this clause specifically excludes temporary buildings, plant, tools and equipment owned or hired by the contractor and any sub-contractor. Where work involves an extension or alterations to an existing building or where the employer chooses to insure the works himself clause 9.1.2.1 should apply. This clause covers exactly the same matters as clause 9.1.1.1 except that where clause 6.3C (IFC 84) is relevant the existing structures are also included and these too are at the sole risk of the employer as regards loss and damage by the 'perils'.

From this it can be seen that the relief from liability which clauses 9.1.1.2 and 9.1.2.2 afford the sub-contractor is restricted and that the sub-contractor still has responsibilities where the risks are not contained within the provisions of main contract insurance clauses. This is illustrated by clause 9.3.1 which makes the sub-contractor responsible for loss of or damage to all materials or goods properly on site for incorporation in the sub-contract works. This is of course subject to clause 9.1 and means that the sub-contract is responsible except for loss and damage occasioned by the perils. Clause 9.3.1 also excludes the materials etc. when they are fully, finally and properly incorporated into the works, and where loss or damage is due to any negligence, omission or default of the contractor etc.

The net effect of these clauses is to make the sub-contractor responsible for a range of risks such as theft, vandalism and conversion. Nevertheless, there is no contractual requirement to insure against these risks though there is a footnote to the contract which suggests that the sub-contractor may consider whether the risks he assumes under clause 9.3 should be covered by insurance.

In addition to these risks the sub-contractor has the risk associated with the loss and damage to his plant or caused by his plant. Clause 11.2 expressly states that this is a matter for the sub-contractor.

It is prudent therefore for such insurance to be effected and this may entail the sub-contractor taking out a contractors' all risk policy (excluding the 'perils') in order to cover the balance of his risk under the sub-contract.

Once materials have been fully, finally and properly incorporated into the works the main contractor becomes responsible for loss or damage except to the extent that such damage is caused by the sub-contractor. The point at which materials and goods are fully, finally and properly incorporated into the works is extremely debatable. It would not seem unreasonable to take these words to mean that the material is properly fixed and nothing remains to be done to it, but even this interpretation could cause problems.

For example, take a sub-contract comprising the supply and delivery of six specialist doors. Can one door only be fully, finally and properly incorporated? It would seem that this can be so. If the door required a lock to be fitted or the bottom to be removed in order that a carpet could be laid, would that door be fully, finally and properly incorporated? It seems in all probability that it would not. The reality of this situation is that should a claim arise the insurers will sort these situations out between themselves.

As work becomes fully, finally and properly incorporated the risk for it is transferred from the sub-contractor to the contractor, save for damage caused by the sub-contractor. There will come a time when all the work is incorporated and at some later point the practical completion of the works will be achieved. As the risk has already been transferred to the contractor there seems little point in clause 9.3.3 which simply confirms this position.

The requirements of clause 9 in no way modify the sub-contractor's obligations under clauses 15.3 and 15.4 in respect of making good defects in the sub-contract works.

Claim for loss or damage caused by 'perils'

Where the main contractor is responsible for insuring the works the sub-contractor is not required under the sub-contract to give any notice to the contractor of damage caused by the perils to the sub-contract works. However, where the sub-contract works are so damaged the sub-contractor will almost certainly inform the contractor because under clause 9.2.1 the contractor is required to pay the sub-contractor the full value of the loss and damage. The full value of damage to which the main contractor is entitled is governed solely by his contract of insurance because under clause 6.3A.4 he is entitled only to the monies received from the insurance.

It has been suggested (see Dennis Turner's *Building Sub-Contract Forms*, page 86, re nominated sub-contractor) that the sub-contractor is only entitled to payment on the same basis as the main contractor. It is submitted that this is not the position because the sub-contractor is not bound by the main contract but by the terms of the sub-contract. Therefore, the value of the loss and damage is to be calculated as though the reinstatement of the damaged works was being done under the instruction issued as to the expenditure of a provisional sum. This means the work will be valued in accordance with clause 16 (see pages 174–79).

It therefore follows that the sub-contractor will be paid on this basis subject to retention in the succeeding interim valuation for the reinstated work.

As already indicated, not everyone would agree with this interpretation so it is clear that the drafting of these clauses leaves something to be desired. The sub-contractor is obliged to reinstate the works because under clause 5.1 he is required to carry out and complete the sub-contract works.

Where the employer is responsible for insuring the works a different position exists in the event of loss or damage caused by the 'perils' to the sub-contract works. The sub-contractor is under a contractual obligation to notify the contractor immediately upon discovery of loss or damage to works by the 'perils'. The notification is required to include the extent, nature and location of this damge.

Clause 9.2.2.1 requires that where clause 6.3B (IFC 84) applies and loss and damage occurs it shall be disregarded in computing any amounts payable to the sub-contractor. Therefore, one treats the sub-contract as though no damage had occurred except for the fact that the cost of the restoration work etc. will be paid for as though it were a variation required by the contractor. The sub-contractor is therefore paid directly by the contractor for the original work and the restored work.

This in effect puts the sub-contractor in the same position as he would be if clause 6.3A (IFC 84) applied. It does not appear from the different wording used in the clauses that this is intended. The different words probably give rise to interpretations to achieve different ends, even if the words used do not fully support such interpretations.

If the employer is insuring the works of alteration or extension then clause 6.3C (IFC 84) will be applicable, and clause 9.2.2.2 provides for a similar situation to clause 6.3B when the damaged work is restored. But additionally, where the work is not required to be restored the employment of the main contractor can be determined and the sub-contractor's employment by the contractor likewise determined in accordance with clause 29 (see page 205).

It should be noted that the references in clause 9.2.2.2 to the main contract conditions and clause 31 are inaccurate.

Policies of insurance

The sub-contractor and contractor each have a right to require the other to provide evidence of insurance to cover any matters covered by the sub-contract or by the main contract as referred to in item 6 of section I of NAM/T. This requirement to produce evidence only applies where the request is reasonable.

It seems unlikely that a request for evidence of insurance cover would be made unreasonably and even where it was whether the other party would see any benefit in contesting that it was given. However, a party may not be satisfied with the documentary evidence provided and clause 10 gives the additional right, so long as it is not exercised unreasonably or vexatiously, to require to have produced for inspection the policy or policies and the receipts of premium paid. The sub-contractor may be required to provide evidence on a periodic basis in order to ensure that appropriate policies of insurance continue to be maintained.

If there is a default by the sub-contractor in respect of effecting the insurance requirements required by the sub-contract then the contractor may himself insure against the risk for which there is default. The default may be in respect of insuring or in continuing to insure as provided by the sub-contract.

The non-production of evidence of insurance when required by the contractor is not in itself conclusive proof that a default has occurred. The non-production of evidence may simply be a technical breach. Therefore, one is not sure at what stage the contractor can assume a default has occurred. It is possible that a sub-contractor fails to produce evidence but has in fact insured accordingly. Can the contractor in this situation insure himself and recover the cost of the insurance from the sub-contractor? It would not seem unreasonable for the contractor to do this but if the sub-contractor has good reasons for not having produced documentary evidence – e.g. lost in the post, appropriate documents not issued by insurance company – then it would seem the contractor could not rely on this clause.

It would be a much easier clause to implement if the default related to the failure to provide documentary evidence rather than the failure to insure.

Where the contractor does insure himself because of a failure on the part of the sub-contractor he may deduct the cost of the insurance cover from any monies due or which become due to the sub-contractor. The contractor must, however, observe the rules relating to set-off under clauses 21 and 22 (see pages 182–190). Alternatively, the contractor may recover the amount as a debt from the sub-contractor.

Although the contractor has a right under the sub-contract to insure himself and to recover the cost, where the sub-contractor is in default, there is no right for the sub-contractor to insure where the main contractor fails to insure his obligation in accordance with the main contract. The reason for this must be that:

(1)   where the main contractor fails to insure the employer himself will take out the necessary insurances.
(2)   there is no sub-contract term that the contractor will insure, but simply a statement as to which main contract condition will apply.
(3)   the impact upon the sub-contractor of the main contractor's failure to insure is relatively small.

Contractor and sub-contractor not to make wrongful use of or interfere with the property of others

It would seem self-evident that contractors and sub-contractors should not wrongfully use or interfere with the property of others and therefore it is perhaps surprising to find a contract clause (clause 26) to this effect:

'The contractor and the sub-contractor respectively and their respective servants or agents or sub-contractors shall not wrongfully use or interfere with the plant, ways, scaffolding, temporary works, appliances or other property belonging to or provided by the other ...'

The significant issue is that there is an attempt to make the sub-contractor responsible for his servants and agents with regard to the wrongful use of plant etc. The clause itself does not entirely achieve this end because it does not expressly impose this obligation on the sub-contractor but rather on the servants and agents (with whom there is no privity of contract). The sub-contractor may well be vicariously responsible for the acts of his servants or agents, and as this is so it is submitted that the clause adds nothing to the legal responsibilities the sub-contractor would normally owe to others.

It should be noted that the clause only refers to 'wrongfully use or interfere'. This recognises the fact that the sub-contractor and/or contractor may well use the plant etc. of others so long as its use is not wrongful and there is no interference. It is therefore extremely important that the sub-contractor and contractor define precisely the facilities and attendance which are available for the use of the other in connection with the sub-contract works. Without such a definition the determination of wrongful use is difficult to establish.

Clause 26 goes on to say:

'... or be guilty of any breach or infringement of any Act of Parliament or bye-law, regulations, order or rule under the same or by any local or other public or competent authority ...'

Once again this part of the clause is not strictly needed from a legal point of view because the sub-contractor is in any event responsible for complying with Acts of Parliament in so far as they affect him. This principle is illustrated by the case of *Street and Another* v. *Sibbabridge Ltd and Another* (1980) where under an implied term the contractor was held responsible for compliance with building regulations.

The clause also contains a proviso which reads:

'... provided that nothing herein contained shall prejudice or limit the rights of the Contractor or of the Sub-Contractor in carrying out their respective duties or contractual duties under the sub-contract or main contract.'

This proviso can only apply to the part of the clause dealing with wrongful use or interference with plant etc. because there can be no question of the contractor or sub-contractor being relieved by a contract term from compliance with statutory regulations unless the statute specifically provides otherwise.

The effect of the proviso would therefore appear to permit either the contractor or sub-contractor to wrongfully use or interfere with the plant etc. of the other if it is used pursuant to their statutory or contractual duties. This raises three questions:

(1)   does this proviso give each carte blanche to use the plant etc. of the other so long as it is in pursuance of his statutory or contractual duties?
(2)   is the proviso making provision only for emergencies which are not provided for in the sub-contract documentation?
(3)   is the proviso permitting the 'wrongful use' as compared with the 'proper use', regardless of whether it is referred to in the sub-contract?

If the answer to (1) above is yes, it would seem that the clause is somewhat tortuously drafted. If the answer to (2) above is yes, this would make more sense as it would enable the plant to be used when it would otherwise be wrongful because the sub-contract did not expressly provide for such specific use. One can hardly accept the possibility that the answer to the third question is yes, because this would surely put an inappropriate interpretation on the word 'wrongful'. As used in clause 26 wrongful must surely mean not provided for in the sub-contract documentation, and not improper use of other people's property in the sense that it may be dangerous.

Amendment 1 issued November 1986 (NAM/SC)

Insurance and related liability provisions

This amendment was issued by the JCT because of continuing concern with regard to the existing indemnity and insurance provisions. A number of these problems have been referred to earlier in this chapter. The amendment is derived from and substantially reproduces the amendments to the insurance and related liability provisions in Amendment 3 issued November 1986 for the Standard Forms of Nominated Sub-Contract (NSC/4 and NSC/4a).

The overall effect of these amendments is that clauses 7 to 11 have been substantially redrafted. The new clause references are used in the following text unless otherwise stated.

Injury to persons and property – indemnity to contractor

Clause 7 now provides for definitions which are to apply to clauses 7 to 11. The definitions included are in respect of:

(1) the contractor or any person for whom the contractor is responsible

which is defined as: the contractor, his servants or agents, or any person employed or engaged upon or in connection with the main contract works or any part thereof, his servants or agents (other than the sub-contractor or any person for whom the sub-contractor is responsible), or any other person who may properly be on the site upon or in connection with the main contract works or any part thereof, his servants or agents; but such persons shall not include the employer or any person employed, engaged or authorised by him or by any local authority or statutory undertaker executing work solely in pursuance of its statutory rights or obligations.

(2)   the sub-contractor or any person for whom the sub-contractor is responsible

which is defined as: the sub-contractor, his servants or agents, or any person employed or engaged by the sub-contractor upon or in connection with the sub-contract works or any part thereof, his servants or agents or any other person who may properly be on site upon or in connection with the sub-contract works or any part thereof, his servants or agents; but such persons shall not include the contractor or any person for whom the contractor is responsible nor the employer or any persons employed, engaged or authorised by him or by any local authority or statutory undertaker executing work solely in pursuance of its statutory rights or obligations.

(3)   terminal dates – which are defined in clause 7.4.

These definitions will naturally aid the interpretation of the contract and not leave it solely to the courts to decide for whom the contractor and sub-contractor should and should not be responsible. Unfortunately, however, any such addition to the contract tends to confuse the administrator of the contract and impair his comprehension. In other words, the clauses are drafted more for the benefit of lawyers than of the building team. This is understandable but it does cause concern to those entrusted to supervise the contract.

Injury to persons

Clause 7.2 alters the existing position in that it now uses the expression 'except to the extent that the same is' instead of 'unless'. The rationale behind this change is that the existing clause (clause 7.1) would prove useless as an indemnity if any part of the damage was caused by the persons specifically excluded by the clause. This new clause is, by the use of the words 'except to the extent that the same is', an attempt to ensure that this is not the case.

The events that reduce the extent of the sub-contractor's indemnity have been widened to include 'breach of statutory duty, omission or default' and these events, together with any act or neglect 'of the contractor or any person for whom the contractor is responsible or of the employer or any person for whom the employer is responsible including persons employed or otherwise engaged by the employer to whom clause 3.11 of the main contract conditions refers or of any local authority or statutory undertaker executing work solely in pursuance of its statutory rights or obligations', are not the responsibility of the sub-contractor.

It would seem that a further definition could have been added in respect of 'the employer or any person for whom the employer is responsible'.

Notwithstanding the above reduction, the sub-contractor's indemnity has been extended by virtue of the definition given to 'the sub-contractor or any person for whom the sub-contractor is responsible' because this now includes 'any person employed or engaged upon ... the works ... or any person who may be properly on the site ...'.

Injury to property

Clause 7.3 provides, subject to clause 7.4 and if applicable clause 9C.1.1, that the sub-contractor indemnifies the contractor in respect of any expense, liability, loss, claims or proceedings in respect of any injury or damage whatsoever to any property, real or personal, including the works, which is due to any negligence, breach of statutory duty, omission or default of the sub-contractor, or any person for whom the sub-contractor is responsible.

The significant differences between this and the existing clause (clause 7.2) are:

(1)   The indemnity extends to include the works. However, there are important exclusions to this which are referred to in clause 7.4.
(2)   The clause attempts to ensure that the sub-contractor is still responsible even though he may only be responsible for part of the damage.
(3)   The sub-contractor is additionally made responsible for breach of statutory duty.
(4)   The reference under the existing clause 7.2 to certain matters being at the sole risk of the employer has been amended but not, it seems, to avoid the consequences of the employer being liable for damage to the works (or existing structure and contents where clause 9C applies) caused by the sub-contractor's own negligence which has created a 'specified peril', e.g. fire.

Exclusions to the extent of liability for injury to property are contained in clause 7.4. This clause reduces the extent of the indemnity by excluding injury or damage to the main contract works and/or site materials, and also where clause 9C applies to injury or damage to the existing structures and the contents thereof owned by the employer or for which the employer is responsible, which is occasioned by a 'specified peril'. This exclusion operates regardless of whether the specified peril is caused by the negligence, breach of statutory duty, omission or default of the sub-contractor or any person for whom the sub-contractor is responsible. But it only operates until the date of practical completion of the sub-contract work or earlier determination of the employment of the contractor under clauses 6.3C.4.3, 7.1, 7.2, 7.3, 7.5, 7.6 or 7.8.

It is interesting to note the use of the word 'works' in clause 7.3, and the words 'main contract works' in clause 7.4. Whether this is intentional is debatable. 'Works' refers to both the main contract and the sub-contract works and this is no doubt intended. However, whether the intention to restrict the exclusion referred to in clause 7.4 to the main contract works is appropriate in the context of this clause is open to question. It would seem that this is the intention, but such a distinction seems rather subtle. One might, perhaps, more readily question whether the employer's insurers should still be responsible for the specified perils even though these are caused by the sub-contractor's negligence.

Insurance against injury to persons or property

Clause 8 is amended in an attempt to make clear the sub-contractor's obligation in respect of what insurance he is required to take out and that such insurance is distinct from the indemnity given under clause 7. Clause 8, therefore, is written without prejudice to the indemnity given by the sub-contractor to the contractor. This new clause 8.1 is now specific in that not only should the sub-contractor maintain such insurance as is specified but he must take out such insurance if he does not already possess the cover required. The intention is the same but the legal interpretation of the earlier words could have proved defective; that is, one was required to maintain the necessary insurance but not to take it out in the first place.

The insurance that is required to be taken out must comply with clause 8.2 (this clause is virtually identical to the former clause) in respect of claims arising out of the sub-contractor's liability referred to in clauses 7.2 and 7.3 as modified by clause 7.4. This means, in general terms, that the sub-contractor is required to insure against personal injury and deaths and injury or damage to property including the works, but not to the extent that the damage is caused by the contractor or employer or those for whom they are responsible; nor to the extent that the damage to the works (or where clause 9C applies to the existing structure) is caused by a 'specified peril'.

Clause 8.1 also restricts the sub-contractor's obligation in that he is not required to take out and maintain insurance for injury and damage to the sub-contract works caused by a risk other than the 'specified perils' up to the terminal date. The terminal date is as referred to in clause 7.4 (see page 117).

However, a sub-contractor may be advised to do so, and the footnote to clause 8.1 also further amplifies the position. It reads:

'The Sub-Contractor has the benefit of the Main Contract Works insurance for loss or damage by the Specified Perils to the Sub-Contract Works but not for other risks e.g. subsidence, impact, theft or vandalism. The insurance to which clause 8.1 refers in respect of injury or damage to property is a third party or public liability policy; such policy will not however give cover for any property such as the Sub-Contract Works while they are in the custody and control of the Sub-Contractor. For this reason the obligation to insure under clause 8.1 is modified by this sentence. As the Sub-Contractor is liable for those other risks if they cause loss or damage to the Sub-Contract Works he may well consider that he needs to take out a Works Insurance to provide such cover which he does not get under the Main Contract Works insurance.'

The new Clause 8.2 differs only in the last sentence. The amount of insurance cover required is now 'the sum stated in the Sub-Contract Documents' as compared with 'the sum stated in NAM/T, Section II, item 2 (or such greater sum as the sub-contractor may choose)'. No reference similar to that in brackets has been retained, but the point is mentioned in the footnotes. However, the reference to sub-contract documents appears to make matters less clear. If it is the intention to leave the published NAM/T as it is this could lead to ambiguity. On the other hand, if NAM/SC is amended then the insertion of the sum required may be overlooked and not included in the sub-contract documents.

Clause 8.3 now makes it clear that the sub-contractor shall not be liable to idemnify or insure in respect of the excepted risks. A definition of excepted risks appears in clause 1.3 and includes:

'ionising, radiation or contamination by radioactivity from any nuclear fuel or from any nuclear waste ...'

The earlier clauses did not make this clear and, although generally any insurance the sub-contractor would have taken out would have excluded these risks, the sub-contractor's contractual responsibility was less clear.

Loss or damage to the main contract works and to the sub-contract works

Generally, the main insurance options have been maintained, but significant changes to format and content have been made. Clauses 6.3A, 6.3B and 6.3C of the main contract now have a corresponding clause in the sub-contract, that is, 9A, 9B and 9C. In addition to this a totally new clause 9 has been introduced to tidy up some very significant loose ends.

Clause 9.1 simply recognises that the main contract clause stated in the sub-contract documents (NAM/T, section I, item 6) will determine the corresponding sub-contract clause which is applicable.

Clauses 9A.2.1, 9B.2.1 and 9C.2.1 each include an exception which relieves the sub-contractor of certain obligations. Clause 9.2 sets out the extent of this exception and reads as follows:

'shall extend to any loss or damage for which either the Employer or Contractor as Joint Insured under the Joint Names Policy ... does not make a claim under that Policy or to the extent that no claim under that Policy can be made because of a condition therein that the insured shall bear the first part of any claim for loss or damage.'

This last reference is of course to any excess that the insured has agreed to take in respect of the cover given.

The main contract now makes provision as to what happens to any insurance monies that are paid out in respect of claims. These monies are now to be paid to the employer, and clause 9.4 of NAM/SC is in recognition of this situation and ensures that the sub-contractor cannot object to this course of action.

Two further clauses appear in this part of the contract, namely clause 9.3 which has the same effect as the existing clause 9.5 and the new clause 9.5 which states that any loss or damage caused by 'specified perils' shall be disregarded when establishing amounts to be paid under the sub-contract. Clause 9.5 now applies to all options whereas previously these words were only included in Clause 9.2.2 which covered only the options under 6.3B and 6.3C. This change will not affect the position as generally accepted but it does clarify the existing position and avoids a possible dispute over the words which do not currently extend to clause 6.3A.

Sub-contract works in new buildings – main contract conditions, clause 6.3A

Where clause 6.3A of the main contract applies, clause 9A of NAM/SC is applicable. This clause now requires that the main contractor shall prior to commencement of the sub-contract works ensure that the all risks policy shall be issued or endorsed to the effect that either the sub-contractor is recognised as insured under the joint names policy or that the insurers waive any rights of subrogation they may have against the sub-contractor until the 'terminal dates'. This position is somewhat different to the position under sub-clauses 9.1.1.1 and 9.1.1.2 because the sub-contractor now either becomes insured under the main contract insurance or is relieved of the consequence of the insurers using any rights of subrogation against him. It is interesting to note that the waiver of any rights of subrogation continues up to and including whichever is the earlier of the 'terminal dates'. Does this mean that once the 'terminal date' has passed any rights of subrogation, even in respect of events before the 'terminal date', can be pursued?

Clause 9A.3 requires that where loss or damage to the sub-contract works or materials occurs before the earlier of the 'terminal dates' the sub-contractor shall upon discovery give written notice to the contractor. This provision has now been extended to all three options whereas previously it did not apply where clause 6.3A was operative. The sub-contractor is also under an obligation to restore and complete the sub-contract works. This always has been the case but was previously included in the general obligations under clause 5 whereas now it is an express provision within the insurance clause. In addition to this the sub-contractor is responsible under clause 9A.2.1 for the cost of restoring the work referred to above except to the extent that the loss or damage is due to:

'one or more of the Specified Perils ... or any negligence, breach of statutory duty, omission or default of the Contractor or of any person for whom the Contractor is responsible or of the Employer or any person employed, engaged or authorised by him or by any local authority or statutory undertaker executing work solely in pursuance of its statutory rights or obligations.'

Where the sub-contractor is not responsible for the cost of restoration etc. by virtue of these exceptions he is still responsible for completing the works in accordance with the contractor's directions. The cost of this work shall be treated as if it were a variation and valued accordingly.

The clause dealing with sub-contract materials or goods that have been fully, finally and properly incorporated into the works, where practical completion of the sub-contract works has not been achieved, has been amended. This clause now makes it clear that the sub-contractor continues to be responsible for loss and damage to these materials or goods, but not if caused by a 'specified peril'. This was previously intended but the wording used was unclear. In addition, clause 9A.2.2 makes it clear that the cost of restoration will also fall upon the sub-contractor to the extent that the loss or damage is caused by the negligence, breach of statutory duty, omission or default of the sub-contractor or any person for whom the sub-contractor is responsible.

It can be seen from the foregoing that the sub-contractor carries risks not covered by the 'all risk' insurance and should consider whether he wishes to take out insurance to cover any risks which may fall upon him.

Once practical completion of the sub-contract works has been achieved the position is that the sub-contractor is no longer responsible for the cost of the loss or damage or for the restoration of the work except to the extent that the loss or damage has been caused by negligence, breach of statutory duty, omission or default of the sub-contractor or of any person for whom the sub-contractor is responsible. This position is stated in clause 9A.5 and is similar but by no means identical to the existing clause 9.3.3.

It should be noted that if a specified peril occurs before practical completion then the sub-contractor is not responsible for the cost of loss or damage. This is so even if the specified peril has been caused by the sub-contractor's own negligence etc. However, after practical completion if loss or damage was caused by a specified peril then to the extent that the sub-contractor is negligent etc. he is responsible for the loss or damage occasioned. Sub-contractors should ensure that their insurance policies cover this risk.

Sub-contract works in new buildings – main contract conditions, clause 6.3B

Where clause 6.3B of the main contract applies, clause 9B of NAM/SC is applicable. Clause 9B is identical in effect to clause 9A, the only difference being that the employer is charged with taking out the insurance and not the contractor. Therefore clause 9B.1 requires that the *employer* arranges that the joint names policy is issued or endorsed as required. In all other respects the format and content of clause 9B are the same as those of clause 9A.

Sub-contract works in existing structures – main contract conditions clause 6.3C

Where clause 6.3C of the main contract applies, clause 9C of NAM/SC is applicable. Clause 9C has the same format and content as clauses 9A and 9B but is not identical on account of the relevance of existing structures.

The differences between clause 9C and clauses 9A and 9B can be summarised as follows:

(1)   Clause 9C.1 has an additional requirement to the effect that the contractor shall ensure that the employer arranges insurance in respect of loss or damage by a 'specified peril' to the existing structures and the contents thereof owned by employer. There is a similar requirement with regard to the issue or endorsement of this policy for the benefit of the sub-contractor.

(2)   Clauses 9C.3.2 and 9C.3.3 are alternative provisions which are dependent upon whether or not the main contractor's employment is determined under clause 6.3C.4.3.

Clause 9C.3.2 provides that where the loss or damage gives rise to the determination of the employment of the main contractor then clause 29 of the sub-contract shall apply as if the employment of the contractor has been determined under clause 7.8 of the main conditions. This replaces a similar clause (clause 9.2.2.2) but it contains a very significant difference in that the original clause referred to the determination of the main contract under clause 28 and determination of the sub-contract under clause 31.

Both of these references were inaccurate and therefore a tidying up of the clause has taken place. It should be noted that where determination of the main contract takes place under clause 6.3C4.3 the provisions of clause 7.7 are applicable except for the words 'and any direct loss and/or damage caused to the contractor by the determination', at the end of clause 7.7.

The sub-contract intends to provide a similar situation in that clause 29 shall apply. This clause is qualified by the reference 'as if the employment of the contractor had been determined under clause 7.8 of the Main Contract Conditions'. This is an error because the reference in that clause is to suspension of the works for a stated period on account of loss or damage occasioned by the 'specified perils'. The appropriate clause is 7.9. Fortunately, this error is not significant because determination under clauses 6.3C.4.3 and clause 7.8 are both governed by clause 7.9.

If the employment of the main contractor is not determined then clause 9C.3.3 is applicable and the sub-contractor is required to restore etc. the sub-contract work in exactly the same way as provided by clause 9A.3.

Clause 9C differs in another way from the original clause because it expressly restricts restoration to the sub-contract works, and not, as might otherwise be implied to the existing structure.

Restoration of work

The main contractor's obligation to restore damaged work now arises as soon as the insurer's inspection of the damage has taken place. But it appears that the sub-contractor's obligation to restore any sub-contract work damaged by the 'specified perils' is not so constrained. Generally, the sub-contractor would await to hear from the contractor that the insurers had inspected the damage before proceeding with the restoration. However, it does seem that the contractor is not under an obligation to provide such information or give any such instructions. The sub-contract only anticipates that the contractor may issue instructions, not that he will. Clearly, it is in the interest of the sub-contractor to check with the insurers that they have carried out their inspection of the damaged work before fulfilling his obligation to restore this work and complete the sub-contract.

Policies of insurance – production – payment of premiums

The new clause 10 is not significantly different from the existing clause in effect, but it is very different in format. The clause is now expressed in four parts.

10.1 sub-contractor to supply documentary evidence as and when reasonably required.

10.2: if there is a default by the sub-contractor.

10.3: main contractor to supply documentary evidence as and when reasonably required.

10.4: if there is a default by the main contractor.

The sub-contractor and main contractor could previously request that documentary evidence be produced to show that insurance had been properly effected. And additionally, they could require to have produced for inspection the policy and premium receipts. The main difference is that previously one was unsure to what insurances the documentary evidence referred as the clause was not precise. The new clauses refer specifically to:

(1) 'Documentary evidence showing that the insurance required under clause 8 has been taken out', and furthermore that it is being maintained by the sub-contactor.

(2) 'Documentary evidence of compliance by the Contractor with the provisions of clause 9A.1 or clause 9B.1 or clause 9C.1.' But this does not apply where the employer under the main contract is a local authority and has opted to use clause 6.3B or 6.3C. Likewise, under the main contract, the contractor has no right to require documentary evidence from the employer in these circumstances.

Where the sub-contractor fails to insure as required by clause 10.1 or the main contractor fails to insure as required under clause 10.3 the other party may then, himself, take out insurance. The cost of this insurance shall be paid by the defaulting party to the party who has now insured. Failing such payment the amount paid becomes recoverable as a debt. The clause does not make express provision to deduct any such sum from any monies which will become due to the party defaulting. This is different from the wording used in the original clause 10.2.

Sub-Contractor's responsibility for his own plant etc.

Clauses 11.1 and 11.2 have been replaced by a single clause 11. Previously the sub-contractor was made responsible for all his own plant, tools, equipment and materials and goods for incorporation in the sub-contract works. In addition he was required to insure against risk except for any loss or damage due to negligence, omission or default of the contractor, his servants or agents or sub-contractors (other than the sub-contractor, his servants or agents).

The position under the new clause 11 is the same except that:

(1) the responsibility (under this clause) does not extend to sub-contract site materials – these of course are covered under the all risks insurance referred to in clause 9.

(2) the exception now only applies to the *extent* that such loss or damage is due to any negligence, *breach of statutory duty*, omission or default (see page 104).

Generally

The amendments to the insurance and indemnity provisions certainly tidy up many of the problems but unfortunately the drafting of the new clauses leaves much to be desired. It now seems that the administrators of the contract provisions will have to rely even more heavily upon specialist advice if they wish to understand what should be done.

Attendance

Under clause 25, which is headed 'Attendance', the sub-contractor is required to clear away his rubbish either to a place provided on site or to a dump. This item is not strictly attendance but is no doubt included under this heading because under other contracts it is usual for the main contractor to have the responsibility for removing the sub-contractor's rubbish. Thus attention is drawn to this different situation.

In addition to this the sub-contractor is required to keep access to the sub-contract works clear at all times and to properly clear up and leave clean and tidy both the sub-contract works and any areas used by him. Any areas would extend to off site areas where such were made available by the contractor for the purposes of performing the sub-contract. See chapter 8 for full discussion concerning attendance and other benefits.

Liability for defects

The liability for defects is dealt with in chapter 9.

Fair wages

Clause 30 requires that the sub-contractor shall in respect of all persons employed by him in every factory, workshop or place occupied or used by him for the execution of this sub-contract comply with the provisions of Supplemental Condition E of the main contract conditions. The wording of this clause raises a number of interesting points.

Firstly, clause 5.7 of the main contract conditions requires that the Supplemental Condition E shall apply only where the employer is a local authority. Therefore it would be reasonable to assume that it was intended that clause 30 of the sub-contract would only be applicable if the employer in the main contract was a local authority. However, clause 30 does not state this but instead requires compliance with Supplemental Condition E. This supplemental condition has as its head reference to clause 5.7, and one may therefore conclude that this is also a part of the conditions and fulfilment of this clause a conditon precedent to its operation. One accepts that this is the intention but it is unclear in law that this would be the case.

Secondly, clause 30 applies to all persons employed by the sub-contractor whether employed in the execution of the sub-contract or otherwise. This wording imposes a very wide obligation upon the sub-contractor and it is doubtful that the reference to 'every factory, workshop ...' is necessary in the light of the reference to all persons under this contract or otherwise.

Thirdly, condition E3 requires that the contractor warrants that to the best of his knowledge and belief he has complied with the general conditions required by the Supplemental Condition E for at least three months prior to the date of his tender for this contract. This raises the question as to whether the sub-contractor has to also warrant in similar terms.

Fourthly, the supplement conditions generally make reference to the contractor and therefore the clause would need to read *mutatis mutandis* with regard to the sub-contractor in order to operate.

These points make the implementation of the supplemental condition cumbersome and may even question its validity in the sub-contract.

Amongst the requirements of fair wages provision is the obligation that a copy of the Supplemental Condition E is displayed at all times in all the places used for carrying out the works. This requirement is frequently not complied with and sub-contractors should take note that such non-compliance may lead to a determination of their employment under clause 27.1.4. However, as it is the contractor who would have to commence the determination process it is in practice perhaps unlikely. Nevertheless, the main contractor is responsible for the sub-contractor's observance of the supplemental condition and if he fails to do so his own employment may be determined by the employer under clause 7.1(d)(IFC).

Strikes: loss or expense

The impact of strikes and the like upon the works is governed by clause 31. This clause ensures that neither the contractor nor the sub-contractor is entitled to loss and/or expense from the other where such loss is caused by strikes etc. affecting their work. This means the parties to the contract take the risk of strikes affecting their work even where the problem is caused by strikes to the other's work.

Notwithstanding this provision the contractor must take all reasonably practical steps to keep the site open so that the sub-contractor can continue working, and the sub-contractor himself must take all reasonably practical steps to ensure that he can carry on working. The provisions of clause 31 are without prejudice to the other rights the parties have under the sub-contract. Therefore, the sub-contractor's and contractor's rights, for instance, to an extension of time caused by a strike would not be affected. Nor would their common law rights under the contract be affected.

The event of a strike is not anticipated as a reason for determination under either the main contract or the sub-contract. However, it is possible that in an extreme case the contract might be considered frustrated in law.

Chapter 8

Attendance and other benefits

Attendance

The main contractor is required under clause 25 to provide the sub-contractor free of charge with:

'... all reasonable hoisting facilities, water, electricity and watching for the purposes of the sub-contract works, space for the storage of materials for use on site and the use of mess rooms, sanitary accommodation and welfare facilities.'

The provision of these items is often referred to as 'general attendance', but the scope of the attendance provided should not be confused with that which is defined in the Sixth Edition of the Standard Method of Measurement of Building Works. The general attendance that is required to be provided under NAM/SC is quite restrictive by comparison.

Other facilities may be provided by the contractor free of charge (see clause 25.3) and these may arise either because the supervising officer has seen fit to include them when completing item 17 of section I of NAM/T or because the sub-contractor has himself requested them by way of variation or addition in item 3 of section II of NAM/T. The sub-contractor should pay careful attention to the items of attendance included, and where he is in any doubt with regard to what constitutes 'reasonable' in accordance with clause 25 he should specifically state what attendance he requires when he completes NAM/T.

The main contractor is required to provide watching, and where he does not do so and the sub-contractor suffers loss the sub-contractor may well attempt to argue negligence on the part of the main contractor and thus benefit from the protection given by clause 11 of the November 1986 JCT Amendment.

It is submitted that the provision by the main contractor of mess rooms, sanitary accommodation and welfare facilities should satisfy the requirements of the Health and Safety at Work Act 1974.

Space only for the storage of materials is required under clause 25 to be provided for the sub-contractor free of charge by the contractor. Any sheds and the like which the sub-contractor requires should be specifically included in item 3 of section II. If such other items are not included here then the sub-contractor is responsible for providing them at his own cost.

These items are referred to in clause 25.4 which extends to items such as workshops and temporary services. The sub-contractor is obliged to provide such items in the place where determined by the contractor, but this is subject to any reasonable objection of the sub-contractor. What constitutes a reasonable objection is difficult to determine but hopefully it should never be necessary to test such words in practice.

The contractor for his part is required to give all reasonable facilities in the erection, maintenance, moving and subsequent removing of these sheds and other items referred to. This clause must be read in conjunction with the attendance item referred to in clause 25.1 and is therefore fairly restrictive in scope. The contractor is required to provide certain facilities to the sub-contractor and these may be required to be provided each time the various sheds etc. are moved.

Scaffolding

Where the sub-contractor erects scaffolding upon the site it may be used by the contractor, his employees and workmen for the purposes of the works whilst such scaffold remains erected. Likewise the sub-contractor, his employees and workmen may use any scaffolding erected by the contractor upon the site. Clause 25.5 provides for this, but this entitlement is not for the exclusive use of the erected scaffolding because the clause provides '... in common with all other persons having a like right ...'.

The clause does not give rise to the use of scaffold by all and sundry but makes it clear that certain rights as to use may be granted by the contractor and sub-contractor alike and therefore the scaffolding may be in use by a number of people at any one time.

NAM/SC does not state who is to provide the various scaffolds for the work, and this is in stark contrast to DOM/1. The latter contract in clause 27.1.2 details precisely the extent of scaffold to be provided by the contractor and the sub-contractor. As the extent to which scaffolding will be provided by the contractor and sub-contractor would normally vary appreciably it has presumably been thought desirable to exclude such a precise requirement. However, in the absence of such a clause the extent of scaffolding to be provided by the contractor and the sub-contractor must be clearly established and set out in NAM/T and other tender documentation.

Although clause 25.5 permits the sub-contractor or contractor to use scaffold erected by the other, it is stated as being:

'... on the express condition that no warranty or other liability on the part of the contractor or the sub-contractor ... shall be erected or implied under the sub-contract in regard to the fitness condition or suitability of the said scaffold.'

This suggests that the use of the scaffold is entirely at the risk of the user and that he should ascertain himself whether it is fit and suitable for the use proposed. Although the user is advised to ascertain whether the scaffold is suitable for the purposes he has in mind, it is submitted that he will not always be at sole risk if the scaffold fails. The failure of the scaffold may be due to negligence on the part of the erector and under section 2(1) of the Unfair Contract Terms Act 1977:

'A person cannot by reference to any contract term ... exclude or restrict his liability for death or personal injury resulting from negligence.'

Under sections 2(2) and 2(3):
'In the case of other loss or damage, a person cannot exclude or restrict his liability for negligence except in so far as the term or notice satisfies the requirement of reasonable.'

'Where a contract term ... purports to exclude liability for negligence a person's agreement to or awareness of it is not of itself to be taken as indicating his voluntary acceptance of any risk.'

Therefore, one can argue firstly that if the clause excludes liability for death or personal injury resulting from negligence then the clause falls foul of the Unfair Contract Terms Act and is void; and secondly, that other loss or damage caused by negligence may not be excluded in so far as the clause is unreasonable.

Bearing this in mind, the net effect of clause 25.5 is that the erector is saying he does not give a warranty that the scaffold erected can be used for the purposes the user has in mind except in so far as the scaffold should satisfy a minimum level of performance and safety.

Benefits under the main contract

Clause 20 provides that:

'The contractor will so far as he lawfully can at the request and cost, if any, of the sub-contractor obtain for him any rights or benefits of the main contract so far as the same are applicable to the sub-contract works but no further or otherwise.'

Initially, this may appear to be giving the sub-contractor rights under the main contract where there is no privity of contract under that contract. The clause should not be viewed in this light because what it seeks to achieve is that where the contractor is himself entitled to a benefit under the main contract he shall seek it and bestow that benefit, to the extent that is relevant, upon the sub-contractor.

If a situation arises which only involves the sub-contractor in some loss then the contractor may not be disposed to pursue any benefit to which he is entitled under the main contract because it will not affect him materially. The existence of clause 20, however, provides a means by which the sub-contractor can secure a benefit under the sub-contract by requiring the main contractor to pursue a benefit to which he is entitled under the main contract and for securing as much of that benefit as is applicable to the sub-contract works. The sub-contractor is required to request such action on the part of the contractor and to stand any costs associated to pursuing the benefit.

Where a sub-contractor requests action under this clause the contractor is required to do all that he lawfully can to secure such benefits. This may simply entail giving notice to the supervising officer who may then act; it may require negotiations; it may require arbitration or even litigation. The cost of such actions fall solely upon the sub-contractor so it would be prudent for the contractor to agree with the sub-contractor the extent to which he will pursue any such claim. The contractor may seek security for the costs involved in such actions but there is no requirement under the contract that such security should be given as a *sine qua non*.

Generally, the majority of matters are dealt with specifically under the sub-contract, and this is fitting bearing in mind that the status of the named sub-contractor is intended to be domestic. Therefore, the occasions upon which the sub-contractor would make reference to clause 20 are limited. The following, however, are examples of when the clause may be used:

(1)   requesting the inclusion of off-site goods under clause 19.4.1.3 (limited use as at discretion of supervising officer).
(2)   inclusion of amounts in respect of fees payable for statutory obligations and the like (see clause 6.1.1 of NAM/SC and clause 5.1 of IFC 84).
(3)   cost of emergency repairs (see clause 6.1.1 of NAM/SC and clause 5.4 of IFC 84).
(4)   inspection of the employer's insurance policy where the work is at the employer's risk (see clause 10.1 of NAM/SC and clause 6.3B.2 of IFC 84).

Although the costs associated with the main contractor securing benefits under the main contractor fall upon the sub-contractor, it may be that the main contractor would be awarded costs in any arbitration and litigation. To the extent that he is awarded such costs the sub-contractor is relieved of his obligation.

Chapter 9

Liability for defects

Practical completion of sub-contract works

Generally, but not invariably, under main contract forms the main contractor is not contractually required to give notice of completion to the supervising officer, although in practice he often will in order to ensure that the supervising officer performs at the earliest moment his obligation of determining if practical completion has been achieved. By contrast, under sub-contract forms the sub-contractor is generally required to give notice of completion to the contractor and NAM/SC is no exception in this respect. Clause 15.1 provides that:

'The sub-contractor shall notify the contractor in writing of the date when in his opinion the sub-contract works are practically completed ...'

For this one can see that a sub-contractor can secure practical completion in advance of completion of the whole works, so long as he gives notice in writing and his work is practically complete. The advantage of securing early practical completion is that the sub-contractor is relieved of certain responsibilities; for example, he is no longer responsible for the effects of frost if these occur after this date. Furthermore, the sub-contractor secures the benefit of an additional 2.5% of the value of work executed once practical completion has been achieved (see chapter 11 for a full discussion on payment).

Once the sub-contractor has given written notice that in his opinion the sub-contract works are practically complete, the date so notified will become the date of practical completion unless the contractor dissents from this opinion. It is up to the contractor, not the supervising officer, to dissent and clause 15.1 gives him 14 days from receipt of the sub-contractor's notice in which to do so. The contractor must register his dissent in writing, and although the clause does not stipulate that it must be communicated to the sub-contractor within the 14 days, it is submitted that it must.

The contractor in deciding whether he should dissent from the notice given should determine whether the works are 'practically complete' because in his notice of dissent he must state his reasons. The clause does not expressly provide that the contractor's reasons for dissent are to be reasonable, and one is therefore led to the view that any dissent will prevent practical completion from occurring. In essence this is correct, but it will be seen that the contractor should act reasonably in this respect otherwise arbitration is a most likely consequence (see also page 138).

In order that the contractor can act reasonably he should have a clear definition of what constitutes 'practically complete'. The ordinary definition of practically is: 'in a practical manner, virtually, almost'. Therefore, one can see that there are two possible interpretations, namely that the works should be complete from a practical point of view or alternatively that the works should be virtually or almost complete.

The main contract form IFC 84 does not use 'practically complete' but simply 'practical completion'. Practical completion could, it seems, also mean either of the two things referred to above. However, it does lend itself more readily to the view that the works must from a practical point of view be complete.

It is most unlikely that the different words used in NAM/SC and IFC 84 are intended, and therefore one would probably wish to ascribe the same meaning to the words used. However, this is not obviously so because the words 'practically complete' in the context they are used are more likely to be interpreted as 'virtually or almost complete'. Support for this interpretation can be found in Donald Keating's *Building Contracts* where he says:

> 'It is thought ... that an architect would not be in breach of duty to his client in issuing the certificate of practical completion where he is reasonably satisfied that the works accord with the contract, notwithstanding that there may be such minor defects *and* (author's italics) always providing that he is satisfied that the retention money will cover their cost *and* (author's italics) that there is no likelihood of the employer suffering loss due to interference with his use of the works while the minor defects are remedied. Further he should obtain a written acknowledgement of the existence of and an undertaking to put right, the defects ...'

Further support can be found in *Nevill (Sunblest) Ltd* v. *William Press & Son Ltd* (1981) where it was decided *inter alia* that the practical completion certificate could be issued where such items as were not complete were considered *de minimus*. This follows the line expressed in *P. & M. Kaye Ltd* v. *Hosier & Dickinson Ltd* (1972) where it was decided that all matters *except trifling ones* (author's italics) must be to the reasonable satisfaction of the architect. It should be noted that under NAM/SC, clause 5.1, the work is only required to be to the reasonable satisfaction of the supervising officer where such approval is so reserved; otherwise, it must be in accordance with the contract documents. This means that the trifling issues may be related to either or both of these standards depending upon which is relevant to the work concerned.

It is submitted that all the above would apply equally to the contractor in determining whether the sub-contract works are practically complete under NAM/SC. From this it can be seen that although it is the contractor who may give notice of dissent to the sub-contractor, the contractor may still nevertheless involve the supervising officer where that part of the sub-contract works has been reserved for the approval of the supervising officer. Therefore, until approval by the supervising officer of work so reserved has been given, the contractor has a good reason for dissenting.

Unfortunately for the sub-contractor, the supervising officer may simply not have viewed the work, for whatever reason, and there is nothing in the sub-contract that enables the sub-contractor to do anything about the situation. This is because the contractor has justifiable reason for dissent and the supervising officer is not a party to the sub-contract. Whether the supervising officer owes a duty of care to the sub-contractor is another matter. One would hope that such situations would not exist, but it is likely that there will be occasions when the supervising officer does not put himself out in this respect, thus leaving the sub-contractor without an effective remedy.

If the main contractor gives notice of dissent, for whatever reason, then practical completion has not occurred for the purposes of the contract. Where this occurs the sub-contractor can do one of three things:

(1) subsequently agree with the contractor what the date of practical completion will be.
(2) refer the matter to arbitration under clause 35.
(3) do nothing and await the practical completion certificate issued by the supervising officer under clause 2.9 of the main contract conditions.

Which of these three courses the sub-contractor would take would obviously depend upon the circumstances. If the sub-contractor accepted the reasons for the contractor's dissent then he would attempt to agree the date of practical completion unless that date were so close to the date of practical completion of the main contract that he could simply accept that this also becomes the practical completion date of the sub-contract works. However, if the sub-contractor disagreed with the main contractor's reason for dissent he might seek arbitration unless, as before, the date of practical completion of the main contract were so close as to make this course not worthwhile. Where arbitration is commenced it will be necessary for the arbitrator to decide whether the reasons given for dissent are reasonable. Therefore, the contractor should only use reasonable reasons for dissent, notwithstanding the fact that the clause does not stipulate that he should.

From a practical stance it can be seen that the contractor has little difficulty in postponing the date of practical completion of the sub-contract works and, in a majority of cases, probably without any ill effects to himself.

Responsibility for defects

Where a defects liability period is contained within a contract this generally enables the sub-contractor to go back and make good his defective work. The defects liability period prescribed in no way restricts the sub-contractor's general liability in law and therefore the contractor remains liable for either six or 12 years as prescribed by the Statute of Limitations Act 1980. The provision of a defects liability period is therefore for the benefit of the sub-contractor in that it gives him a right to make good his own defects rather than allowing the contractor to do the work himself and then charge the sub-contractor for that work.

Under clause 15.3 the sub-contractor accepts a similar liability to that of the main contractor under the main contract to remedy defects in the sub-contract works. The sub-contractor is therefore under an obligation to remedy defects in the sub-contract works up until the expiration of the defects liability period prescribed in the appendix to clause 2.10 of the main contract.

Under JCT 80 the main contractor is only permitted to make good defects etc. which appear within the defects liability period. This, it seems, reduces the benefit of the clause to the main contractor. IFC 84, however, has amended wording which simply refers to 'appear' but by inference could be said to be saying the same thing as JCT 80. By contrast, NAM/SC enables the sub-contractor to go back and make good all defects, shrinkages and other faults due to materials or workmanship not in accordance with the sub-contract which may exist and which may have appeared at any time up until the expiration of the main contract defects liability period.

Strictly, the sub-contractor is only able to go back and make good work which is not in accordance with the contract, and this raises a particularly interesting issue with regard to shrinkages. If the shrinkage would occur in any event when using the materials and workmanship prescribed in the contract documents then this is something for which the sub-contractor cannot return to remedy. Is the sub-contractor, nevertheless, still liable for the shrinkage? It seems not under this contract, because this event does not constitute a breach of contract and it would seem the sub-contractor is quite entitled not to go back on site to rectify and therefore does not require the ability to do so. Notwithstanding the legal position, most sub-contractors are prepared to remedy such shrinkages for the sake of goodwill and because of the difficulty of deciding when a shrinkage is naturally occuring or otherwise.

Clause 15.3 also makes it clear that the sub-contractor is responsible for damage due to or caused by frost occurring before the date of practical completion of the sub-contract works. It is axiomatic, therefore, that once the sub-contract works have reached practical completion the sub-contractor is no longer responsible for damage caused by frost. The contractor then becomes responsible in this respect until the main contract works reach practical completion. Therefore, the main contractor may find himself liable for frost damage to the sub-contract works but unable legally to pass it on to the sub-contractor. However, this will not necessarily stop him from trying.

The sub-contractor has the benefit of being able to go back to make good defects, but naturally at no cost to the contractor. However, the sub-contractor may be denied the opportunity to make good his defects because the supervising officer may decide that such defects are not to be made good. This situation is provided for in clause 15.4 which states:

'where under clause 2.9 of the Main Contract Conditions, the Contractor is liable to make good defects, shrinkages or other faults in the Sub-Contract Works but the Architect/Supervising Officer otherwise instructs and makes an appropriate reduction ... then the

Contractor shall instruct the Sub-Contractor not to make good defects, shrinkages or other faults and an appropriate, deduction shall be made ...'

At first glance this is a strange clause in that it denies the sub-contractor the right to make good his own defects and to put the work into the condition for which he originally contracted. Generally, clause 15.4 will not be implemented. However, it does enable the supervising officer to avoid having the sub-contractor back on site in circumstances where it is perhaps inconvenient or best avoided – for example, where his presence may disrupt the employer's use of the building or where the supervising officer has lost faith in the sub-contractor's ability to perform the making good required.

Where it is decided that the making good should not be executed an appropriate reduction will be made to the sub-contract sum by the contractor. The sub-contractor should ensure that if any such deduction is made it is in respect of defects for which he is responsible and the deduction made is a reasonable one. What constitutes an appropriate reduction in these circumstances is unclear.

Presumably it includes the cost of making good the defects, but whether it is intended to cover other losses such as disruption to the use of the building is doubtful. The fact that the sum does not necessarily cover these latter items does not limit the contractor's claims for breach of contract (*Nevill (Sunblest) Ltd* v. *William Press & Son Ltd* (1981)).

The cost of making good will be assessed, and it seems that any sub-contractor suffering a deduction in this way will claim that the amount is excessive and that he could have executed the work far more cheaply if he had been permitted. There is nothing, it seems, that restricts the cost to what it would have cost the sub-contractor but the cost deducted must reflect what would be a fair and reasonable charge by sub-contractors at large for such work. This could of course mean that in extreme cases the deduction actually exceeds the amount of the original sub-contract sum.

Under the main contract the supervising officer is responsible for producing the schedule of defects and notifying the contractor not later than 14 days after the expiry of the defects liability period. The contractor upon receipt of this schedule should notify the sub-contractor because under clause 15.3 the sub-contractor is under a similar obligation, but the contractor need not restrict himself to this list. The sub-contract does not prescribe a time limit in which the main contractor should notify the sub-contractor what defects are existing, so it will be implied as a reasonable time.

Upon satisfactory completion of all the defects – this means those of the main contractor and the sub-contractor alike – the supervising officer will issue a certificate to that effect. This certificate is generally known as the Certificate of Completion of Making Good Defects, but it is not expressly referred to in these terms in IFC 84. The contractor is not involved in the issue of any certificate of similar effect to the sub-contractor and no separate certificate is envisaged in respect of making good sub-contract defects.

Chapter 10

Time

Generally

This chapter deals with the specific matters of time which relate to sub-contract works. It brings together those matters which are specifically concerned with time, such as commencement, completion, delay in progress, extensions of time and the consequences of delay. The consequences of delay include consideration of the damages provisions and how the loss and expense provisions become operative. However, the main discussion on loss and expense is dealt with in the chapter on payment (chapter 11).

Commencement of the sub-contract works

Most main building contracts now recognise the need for and importance of stating a precise date for possession of site, thus providing the contractor with a firm date both for establishing his tender price and for programming the works. Any change to this date following the signing of a contract has significant consequences, and the contractor has available certain remedies under the contract which generally ensure that he at least will not lose out as a consequence of the work being delayed for this reason.

By contrast, many sub-contracts are entered into without the precise date for commencement having been established. The main reason put forward to support this policy is that there is no way that the main contractor can tie himself to a specific commencement date, bearing in mind all the uncertainties of construction works and the vagaries of the weather. This means that the sub-contractor is put at greater risk in this respect than the main contractor, because when he contracts he does not know when he will be starting and finishing his works. Let us consider what the sub-contractor commits himself to under NAM/SC in respect of time and what risk he carries as a consequence.

When a sub-contractor tenders he will be given certain time related information and this will be set out in NAM/T. This information should include:

Section I

Item 6 – date of possession of main contract works.
Item 6 – date for completion of main contract works.
Item 8 – order of works if affected by the employer's requirements.
Item 10 – new dates for possession of completion where these have been altered from those stated in item 6.
Item 15 – dates between which it is expected that the sub-contract works can be commenced.
Item 15 – period required by the supervising officer to approve drawings after submission.

The sub-contractor will also be able to state in item 1 of section II the periods he requires for submission of drawings, execution of works off site, notice to commence work on site and execution of sub-contract works on site.

Taking all these matters into consideration, the sub-contractor will know no more than that the work may be commenced at some time during the main contract period, or indeed thereafter if an extension of time is in operation. For the reason already stated the sub-contractor is not given a precise commencement date. The sub-contractor's contractual obligation with regard to commencement is contained within clause 12.1, in that:

'The sub-contractor shall carry out ... in accordance with the NAM/T Section I item 15 and Section II item 1 and reasonably in accordance with the progress of the Works, subject to receipt of the notice to commence work on site as stated in NAM/T Section II item 1 and to the operation of clause 12.'

Although there is a side note to item 15 which states that the actual date or dates for commencement of the sub-contract works should be settled by the contractor and sub-contractor, there is no specific provision within NAM/T to insert such dates. Therefore, if a sub-contractor tenders on the basis of NAM/T he may find himself locked into a contract with the main contractor where no such date exists. This can occur because once a tender has been made it can be accepted by the contractor without further discussion on this point, and it seems that most contractors would probably see this as advantageous.

Where a sub-contractor is named in the tender documents the contractor is required under the main contract to enter into contract with the named sub-contractor within 21 days. Therefore, it can be seen that the sub-contractor can be bound at an early stage to execute work at some time in the future so long as the timing is in accordance with the wide time framework.

Fixing the commencement date

The commencement date for the sub-contract works may be fixed by specifying in the contract or at some later stage, dependent upon the factors already referred to. However, whichever way is adopted the sub-contractor can expect and is entitled to receive 'notice to commence work on site' in accordance with item 1 (3) of section II of NAM/T. It is therefore apparent that even where a commencement date is fixed in advance the contractor has the opportunity to notify a new date and to grant an extension of time. Notwithstanding this, it is still advantageous for the sub-contractor to have a precise date for commencement stated in the contract, because although this date may vary it may enable the sub-contractor to recover loss and expense in respect of some types of delay. The alternative allows the contractor much greater freedom and at no cost.

If the date for commencement of the sub-contract works is to be specified in the contract it should be stated in item 1 of section II of NAM/T; otherwise, the validity of any such date would be put in doubt owing to the priority of sub-contract documents.

The notice given by the contractor to the sub-contractor to commence the sub-contract works must be given in accordance with the time specified in item 1(3), section II, of NAM/T. This time period will enable the sub-contractor to organise himself (including executing any off site work) in advance of the date when he is required to commence and from which the sub-contract duration will run.

Example:

Notice to commence work given as	3 May 1987
Notice required by section 1(3)	three weeks
Commencement date, therefore, is	24 May 1987

Example:

Date for commencement stated in NAM/T	12 May 1987
Notice to commence work given on	3 May 1987
Notice required by section 1(3)	three weeks
Commencement date, therefore, is	24 May 1987

It can be seen that regardless of whether the commencement date is stated in the contract documents it is the notice to commence work which is instrumental in fixing the actual sub-contract commencement date. Although the notice to commence work will generally determine the commencement date, is this the case where the sub-contractor and contractor agree to commence in advance of this date? The answer, it seems, is no, and the date of actual commencement on site is thus taken to be date of commencement. The period of notice would in these circumstances be waived, and where this does occur the contractor would be advised to record the actual date of commencement by the sub-contractor on site.

Delay to commencement date

Even though it is the notice to commence work which enables the sub-contractor to start work, where the date is later than the date stated in the contract he is likely to be entitled to an extension of time and possible loss and expense.

Where no such date is fixed a different situation prevails, because there cannot be a delay. However, once the notice to commence work is given any subsequent delay to the date of commencement which is not the responsibility of the sub-contractor may give entitlement to an extension of time and possible loss and expense.

Completion of the sub-contract works

The sub-contractor is obligated to carry out and complete the works, subject to clause 12, within the duration which is stated in item 1(4), section II, of NAM/T. The time runs from commencement date which is the date stated in the notice to commence work and as referred to above. One should note that NAM/T refers to the date stated in the notice and not the date of the notice. These two dates are different and should not be confused.

Delay to progress or completion of the sub-contract works

Because the obligation to complete is subject to clause 12, the sub-contractor may be entitled to an extension of time when he suffers delay. But regardless of whether a sub-contractor is entitled to an extension of time he is obliged under clause 12.2 to give notice of any delay that has occurred or is likely to occur to the progress or completion of the sub-contract work. Once the sub-contractor recognises a delay he must, regardless of whether it will affect the completion date, give notice forthwith to the contractor. This notice must specify the cause of the delay. The contract requires that the cause of the delay shall be given 'in so far as he is able', but it is difficult to imagine circumstances when he would not be able to specify a cause, because not being able to is itself an admission of poor management.

If the delay to which the notice refers is caused by

'any act, omission or default of the contractor, his servants or agents or his sub-contractors, their servants or agents (other than the sub-contractor his servants or agents) or by any of the events in clause 12.4'

and this delay has caused or is likely to cause a delay to the sub-contract works beyond the completion date or previously extended completion date then an extension of time shall be granted. That is so long as the sub-contractor has constantly used his best endeavours to prevent delay and to do all that is reasonably required by the contractor to proceed with the sub-contract works. There is in this clause nothing that requires the sub-contractor to overcome the delay once it has occurred unless it is to lead to another separate delay, and nothing that requires him to accelerate the works.

The events referred to in clause 12.7 are very wide ranging and cover most of the familiar issues contained in DOM/1, but there are a number of notable differences. NAM/SC does not include the following:

(1) delay on the part of the nominated sub-contractors or suppliers (DOM/1, clause 11.10.7). This is simply because under IFC and NAM/SC there are no nominated sub-contractors or suppliers. However, there are the named sub-contractors, and delay on the part of these is a notable exception to the relevant events included in both the NAM/SC and IFC contracts. Therefore, the sub-contractor and contractor alike cannot secure an extension of time

where a named sub-sub-contractor or named sub-contractor causes delay through being dilatory in carrying out the works. This is in line with the philosophy that the use of named sub-contractors is domestic in nature and this issue should be a matter between those directly concerned. However, if the delay which a named sub-contractor suffers is a relevant event the main contractor can secure an extension of time for this as though it were a delay that he himself was suffering.

(2) delay caused by the United Kingdom Government exercising a statutory power which affects the works (DOM/1, clause 11.10.9). Presumably, the spectre of the three day week in 1973 is somewhat distant and therefore the loss of such a clause is easily accepted. There is nothing to say that a similar event will not occur. If it does, the delay it causes would be at the risk of the sub-contractor unless it could be shown to be force majeure or lead to frustration of the contract.

(3) delay caused by the inability to secure labour or materials. DOM/1, clauses 11.10.10.1 and 11.10.10.2 are optional in NAM/SC and their adoption or otherwise has no impact upon the operation of the fluctuation provisions.

On the other hand, NAM/SC includes in clause 12.7.13 for the deferment of the employer in giving possession of the site. This refers to the possibility of a delay to the possession of the site by the main contractor, thus causing the sub-contractor a delay to the date upon which he would otherwise commence. No such clause is included in DOM/1.

As one can see from the discussion on the commencement date (see pages 142–145, unless the commencement date is fixed in advance of the deferment the fact that there is non-possession of site is irrelevant from the point of view of an extension of time for the sub-contractor. This is because in the absence of a fixed commencement date the contractor already has the ability to overcome such a delay simply by notifying a commencement date which takes account of any such delay.

With regard to loss and expense associated with delay caused by deferment of the site, a different position may prevail. It may be argued that any delay in possession of the site caused by the employer is automatically considered a delay in the progress of the works. However, this is not a sound argument because a deferment of site possession may in fact have no impact upon the progress of the sub-contract works, and without a fixed commencement date it is unlikely that much benefit will be gained by a sub-contractor from these clauses.

Clause 12.7.13 concerns itself only with deferment of the employer giving possession under the main contract. It does not refer to delay to the commencement of the sub-contract works which has been caused by the contractor. This would be a matter to consider under clause 12.2 as an act, omission or default.

Extension of time

If the sub-contractor has suffered delay of a type referred to in the contract, has done his best to prevent that delay and has fulfilled all the requirements with regard to the giving of notice and supporting information then the contractor is required to grant an extension of time. The contractor is required to grant an extension of time in writing as soon as he is able to estimate the length of delay beyond the completion date.

When granting an extension of time, two significant points arise. Firstly, the contractor is required to estimate the length of delay, not to know as a fact the actual length of delay; therefore he should not postpone the granting of an extension until he has this knowledge. Secondly, the contractor is required to estimate the length of delay *beyond* the sub-contract completion date, not the length of delay actually occasioned.

In practice these are often taken as the same thing, but this is erroneous. For instance, if a sub-contractor is working ahead of programme and then suffers delay but can nevertheless still finish within the original completion date, no extension of time need be granted.

It was decided in *Miller* v. *London County Council* (1934) that, since the power to extend time had not been properly exercised, the employer lost the benefit of such a clause, i.e. to keep liquidated damages alive. Although other cases such as *Amalgamated Building Contractors Ltd* v. *Waltham Holy Cross Urban District Council* (1952) have shown that extensions of time may in certain circumstances be granted some length after the completion date, the *Miller* case illustrates the need to consider the extension of time at the appropriate moment.

Under many other forms of contract, and particularly JCT 63, it has been a matter of contention whether an extension of time for certain events can be granted once the due completion date has passed. NAM/SC, along with JCT 80 and IFC 84, takes account of this situation and provides in clause 12.3 a means which enables extension of time to be granted for certain specified events even though the date for completion has passed.

Clause 12.3 reads:

'If any act, omission or default ... or an event referred to in 12.7.5 to 12.7.8, 12.7.11 or 12.7.13 occurs after the expiry of the periods stated ... (or after the expiry of any extended period ...) but before practical completion ...'

Therefore, if these delays occur the contractor is still able to grant a valid extension of time. The events referred to cover instructions on behalf of the supervising officer, lack of information which should be supplied by the supervising officer and certain matters under the control of the employer.

This clause addresses one specific aspect of the problem, but one is still left with difficulties. What about the other delays not covered by this clause? These are excluded because the main problem is concerned with the right of the contractor to grant an extension for matters for which he is in default or responsible.

Consider a variation instruction and exceptionally adverse weather conditions both causing delay after the due completion date. It is now clear that an extension of time can be granted in respect of the delay caused by the variation. It also seems, following the comments of Lord Denning in *Amalgamated Building Contractors Ltd* v. *Waltham Holy Cross Urban District Council* (1952), that an extension could be granted for the delay caused by the exceptionally inclement weather. With respect, it is difficult to agree that this latter issue should be the law for the following reasons. Once the contractor has failed to achieve the completion date, he is in default. If a variation is then issued he can secure an extension of time. However, if after completion he also suffers other delays not connected to the variation then surely these are matters which fall upon the contractor. The further delay is a direct consequence of his own default. Where the other delay is connected with the fact that there is a variation a different position may prevail.

It does seem unreasonable that this delay for whatever reason should not be taken into account in granting an extension of time, and it is probably this type of matter which Lord Denning really had in mind. However, even in this type of situation it could be argued that although the further delay is connected to the variation it is not caused by the variation and therefore does not fall under the relevant event. This, it seems, would be a very harsh interpretation of the provisions.

Reviewing extensions of time

Provision is made in clause 12.4 for the granting of extensions of time by the contractor at any time 'in accordance with the provisions' and is an optional means available to the contractor for granting an extension of time. The intention of the clause is twofold. Firstly, it enables the contractor to grant an extension of time on his own initiation. A notice by the sub-contractor is not a condition precedent. Secondly, it enables a review of extensions of time previously granted.

The clause is seldom going to be used for the first purpose because the majority of contractors will, it seems, rely upon the sub-contractor to give notice of delay. As regards the second point, this will enable a review similar to that which will be undertaken by the supervising officer in respect of the main contract works. There is in IFC a time scale for this review but under NAM/SC it can take place at any time. No doubt the contractor will, where appropriate, await the supervising officer's review before reviewing the sub-contract extensions – that is, if the contractor is inclined to perform a review at all. In any event, such a review of the extension of time cannot reduce any extension previously made. This is regardless of whether work has been omitted since the extension was granted.

The position under NAM/SC and IFC is therefore different from that adopted by JCT 80. This means that the first extensions of time made will almost certainly be on the mean side.

Failure to complete sub-contract works on time

Once the commencement date has been established, the date for completion can be calculated by reference to item 1(4) of section II. The sub-contractor is then obligated to complete by this date unless he can secure an extension of time or alternatively can show that because of the contractor's default the completion date no longer applies.

The situations where the sub-contractor can show that he is no longer bound by the completion date because of the contractor's default are somewhat limited. However, it may occur where the contractor is unable to grant an extension of time in respect of a matter for which he is responsible or where the contractor clearly shows that he does not intend to comply with the extension of time provisions. In the fairly rare circumstances of the completion date being set aside, time is said to be 'at large'. The effect of this in some contracts is extremely significant because without a fixed date for completion liquidated and ascertained damages cannot operate. This does not necessarily mean that the defaulting party will escape all liability, because a reasonable time for completion will be substituted and unliquidated damages (as much as can be proved) may be recoverable. But clearly the onus has changed.

Interestingly, NAM/SC does not provide for liquidated and ascertained damages, and any damages caused by delay to the completion date will be recoverable by the contractor as unliquidated damages. Therefore, the significance of 'time at large' in this contract is reduced. The main advantage of keeping the extension of time provision alive in these circumstances is that it does establish a date from which damages are calculated, rather than it being left to a reasonable time for completion. These may, of course, be seen in the last resort as the same thing.

Much discussion about delay on building contracts revolves around the ability or otherwise to grant an extension of time. This is because it is not only important for the contractor and sub-contractor to secure more time to complete and avoid damages for delay, but also for the employer to keep alive any liquidated and ascertained damages that may be included in the contract. If liquidated and ascertained damages cannot be kept alive and time becomes 'at large' then the party suffering damage on account of the delay must prove the actual damages incurred as a consequence of not completing within a reasonable time. A reasonable time would in the last resort have to be established by the courts. Therefore, it is important that where liquidated and ascertained damages are applicable they are kept alive, and for this to be assured extensions of time provisions must be available for events which are the responsibility of the employer (the employer in this sense is the main contractor in a sub-contract) and the provisions must be operated appropriately.

As already mentioned, however, NAM/SC does not provide for liquidated and ascertained damages for delay on the part of the sub-contractor, and therefore this type of discussion, which features in virtually all the legal text books, is largely irrelevant because the contractor has, in any event, to prove his damage. There is nevertheless still an important difference between 'time at large' and the normal situation provided for under NAM/SC. Although both situations are concerned with unliquidated damages the date from which they are calculated is likely to be set by the courts where time is at large and by the contractor where the extension of time provisions can be operated. Therefore, it is important to make provision for extension of time and to operate the clause accordingly, but the consequences of not doing so are less significant where liquidated and ascertained damages are not applicable.

Where the sub-contractor fails to complete the sub-contract works within the specified period or any extension of that period the contractor must notify the sub-contractor in writing. The contractor is required to give this notice 'within a reasonable time of the expiry of that period or those periods'. There is no express provision in this contract for either liquidated or unliquidated damages, and therefore the only damages that can be recoverable are unliquidated damages which flow from the breach of contract – that is, not completing the sub-contract work on time.

By comparison, although DOM/1 does not make provision for liquidated damages it specifically refers in clause 12.2 to '... any loss or damage suffered or incurred by the contractor and caused by the failure of the sub-contractor as aforesaid'. However, it should be noted that clause 14.3 may be used to cover this situation, and this is discussed on pages 156–157.

Clause 13, NAM/SC, appears to make the notice by the contractor to the sub-contractor a condition precedent to the recovery of unliquidated damages on account of the breach. Does this mean that there can be no breach until the contractor confirms that the sub-contract work should have been complete? Under clause 12.2 of DOM/1 it is specifically provided that 'on receipt of the notice ... the sub-contractor shall pay or allow ...', and therefore it is clear that until such a notice is given no right to damages is intended. The position under NAM/SC is less clear because nothing is expressed.

In order that the contractor can issue such a notice he should have considered all the extensions of time to which the sub-contractor may be entitled. This he can do under clause 12.4, but should he fail to do so at this time he may still make a further extension of time at a later date, having previously issued a notice that the sub-contract works should have been complete. Where such an extension is granted after the notice has been given the notice is deemed to be cancelled. There is no reference to the ability to issue a further notice at some late stage, but likewise there is nothing that restricts this action.

Where delay occurs to the sub-contract and the contractor gives notice that the work should have been completed he will no doubt wish to recover his damages. These he will have to establish and may attempt to set-off against sums otherwise due to the sub-contractor. Such claims are dealt with in detail in chapter 12.

If the notice by the contractor is a condition precedent then it does seem that the sub-contractor has a defence against damage for delay if such a notice is not given. One cannot be sure how the courts would view the absence of this notice, but it does seem on balance that the right to recover damages might well be lost. The contractor has a reasonable time in which to issue the notice. This can only be interpreted as within a reasonable time after expiry of the completion date, and what consitutes a reasonable time is of course open to debate. There is little doubt that once the contractor has made up his mind that he wants compensation for the sub-contractor's delay a notice will be forthcoming, albeit late. It would be a brave sub-contractor who would rely entirely upon the argument that the notice is out of time and therefore no damages can be recovered.

Liquidated or unliquidated damages

Building main contracts generally provide for liquidated and ascertained damages in the event of delay, whereas building sub-contracts often favour unliquidated damages. This does not mean that sub-contracts cannot provide for liquidated damages. It is not difficult to provide for them where a preference is shown. However, unliquidated damages are generally favoured because contractors argue that delay occasioned by a sub-contract may have varying consequences and is dependent upon the progress and programme of the main contract works. Unlike the building employer who sees all delays resulting in non-possession of the work, the contractor sees delays by sub-contractors affecting him in a variety of ways.

The disadvantage of unliquidated damages is that they have to be proven and cannot be automatically deducted from sums otherwise due because they are of no pre-determined amount. They may also affect a sub-contractor's tendering strategy in that the possibility of unliquidated damages amounting to thousands of pounds may deter a sub-contractor who is only involved with a relative minor sub-contract. The advantage of unliquidated damages is that they do reflect the true or actual loss occasioned by the delay as compared with an estimate of the loss to be suffered at some future date.

Although a sub-contract may rely upon unliquidated damages for delay caused, it is possible that a part of those damages will include liquidated and ascertained damages payable by the contractor to the employer; hence, the need to give notice to the sub-contractor of what these damages will be (see item 6, section I of NAM/T). This situation may arise where the delay in the sub-contract works causes not only a delay to the contractor's progress but also a delay to the completion date.

There is a potential problem with regard to the liquidated and ascertained damages included in the main contract appendix. This is because item 14, section I, of NAM/T states the main contract appendix and entries therein will, where relevant, apply unless otherwise specifically stated here. It is therefore possible to argue that the liquidated damages inserted are equally applicable to both the main contract and sub-contract alike and that no other damages are recoverable. This, however, is surely not the intention. The liquidated damages stated are intended to give notice of one element of the unliquidated damages recoverable by the main contractor from the sub-contractor. But in order to be certain that they operate in this way an entry should be made in item 14, section I, of NAM/T to this effect.

Disturbance of regular progress of sub-contract works

Where the progress of the sub-contract works is delayed by one of the relevant events and this event has also caused a delay to the sub-contract completion date, the sub-contractor will naturally seek an extension of time. He may also in certain circumstances seek loss and expense for the delay he has suffered. But what of the situation where a delay to the progress of the sub-contract is occasioned but the sub-contract completion date can still be met? Here again, if the delay is one of those matters referred to in clause 14.2 then the sub-contractor can seek his loss and expense. It is immaterial that the delay has not caused a delay to the completion date.

Clause 14 provides a mechanism which enables the sub-contractor to recover loss and expense occasioned by specified disturbances of the regular progress of the sub-contract works. The difficulty for the sub-contractor and main contractor alike is determining that a disturbance has occurred. A disturbance of regular progress is generally taken as synonymous with a delay. Although a disturbance may arise which does not cause delay, a disturbance of regular progress will always cause a delay. The delay may, however, manifest itself through a reduction in productivity or require re-programming, i.e. one event is delayed in preference to another in order to reduce the consequential effect. Therefore, in order that one can determine whether a disturbance of the regular progress has occurred one must:

(1)   have a programme of the sub-contract works.
(2)   be able to monitor and record the actual progress of the sub-contract.

If the sub-contractor identifies a delay to the commencement or to the regular progress of the sub-contract works then the sub-contractor may seek loss and expense in respect of the event. The matters included by clause 14.1 are covered by the words:

> '... due to the commencement or regular progress of the sub-contract works or any part thereof having been materially affected by any act, omission or default of the contractor, his servants or agents, or any sub-contractor his servants or agents (other than the sub-contractor ...) or is materially affected by any one or more of the matters referred to in clause 14.2 ...'

In order that the sub-contractor can secure loss and expense due to the commencement being delayed there must be an agreed commencement date. As already discussed, this will frequently not exist until very close to when actual commencement takes place. This situation therefore restricts substantially the benefit of the clause.

It can be seen from clauses 14.1 and 14.2 that the disruption can occur on account of:

- main contractor's act, omission or default
- other sub-contractor's act, omission or default
- supervising officer's default and specified acts
- employer's default
- sub-contractor's own delay where a valid right to suspend exists

Should the sub-contractor make a written application to the contractor within a reasonable time of it becoming apparent that the sub-contractor has incurred or is likely to incur direct loss and/or expense, then if this is not provided elsewhere in the contract an agreed amount will be added to the sub-contract sum. The sub-contractor may be required to provide supporting evidence to his application, and if he fails to do so there is no obligation upon the contractor to agree the claim.

In this respect it should be noted that clause 14.4 provides that the provisions contained in clause 14 are without prejudice to any other rights that may be possessed. This means that the sub-contractor may pursue his claim by other means, thus avoiding the procedure laid down by clause 14. Because of this it would probably be in the interests of the main contractor to agree an amount of the loss and expense, notwithstanding the fact that the sub-contractor has provided insufficient details. Clearly, however, it is in the interests of the sub-contractor to provide all relevant details in support of his claim.

Disturbance of regular progress of the main contract works

Claue 14 deals not only with the rights of the sub-contractor should he encounter delay to his work but also with the contractor's rights against the sub-contractor where the latter causes delay to the main contract works. Clause 14.3 deals with this matter and provides for the sub-contractor to pay loss and expense where he delays the main contractor. In order that the main contractor can secure a claim, he must show that 'the regular progress of the works is materially affected by any act, omission or default of the sub-contractor, his servants or agents', and, also make a written application within a reasonable time of the material effect becoming apparent.

It is important to note the use of different words in clause 14.3 as compared with clause 14.1. The former clause only requires an application to be made within a reasonable time of the event which has materially affected the main contract works, whereas clause 14.1 requires an application to be made within a reasonable time of incurring or being likely to incur direct loss and/or expense.

Clause 14.1 requires an element of foreseeing, clause 14.3 does not. Such a distinction is dubious: surely, if a sub-contractor is expected to see a likely loss and expense situation arising then why not the contractor also? If clause 14.3 was envisaged only to deal with delay to the regular progress of the works caused by the late completion of the sub-contract works then the use of different wording can be justified. But this surely cannot be the case because delay of the sub-contract works is dealt with specifically under clause 13. Clause 14.3 must be intended to cover delays to the main contract works caused by the sub-contractor failing to perform to his programme but not necessarily being late with regard to completion.

The problem for the contractor in securing loss and expense is obvious because here he is required to show a programme to which the sub-contractor is contractually bound and with which the sub-contractor has without just cause failed to comply. In other words, the main contractor is required to support his application by providing such information as may be reasonably required by the sub-contractor. If no detailed programme has been agreed it seems that the main contractor's chances of securing a successful claim will be greatly reduced. In the absence of such a programme (see clause 12.1) the main contractor would be relying solely upon the implied term that the sub-contractor would proceed to carry out the works regularly and diligently.

Chapter 11

Payment

Sub-contract sum

The amount of the sub-contract sum is to be stated in Article 2 of the Articles of Agreement which are contained in NAM/T. The sum is stated to be exclusive of value added tax, and any tax payable under clause 17A or 17B is to be in addition to the sub-contract sum.

NAM/SC, unlike NSC/1 and DOM/1, does not provide for the alternative of a tender sum. The use of this alternative in the other forms is to be adopted when the works are subject to complete remeasurement, and therefore as NAM/SC does not provide for this alternative it can be seen that the sub-contract work is intended to be firm in design. The sub-contract sum payable is, however, still adjustable in accordance with the conditions of contract, but is intended to be adjusted solely on an 'add and omit' basis and not by complete remeasurement.

Clause 4 makes reference to the adjustment of the sub-contract sum by the addition or deduction of amounts which are provided for in the sub-contract conditions, and also provides for when this adjustment is made.

Quality and quantity

The quality and quantity of work included in the sub-contract sum is dealt with under clause 3. This clause sets out what documents define the sub-contract sum, and because the Intermediate Form of Contract is intended to be used with a variety of documents, a number of options are included.

Without quantities (clause 3.1)

Where there are no bills of quantities used and where there are no quantities set out in the numbered documents, the quality and quantity of work included in the sub-contract sum is that which is described in the sub-contract documents when taken together. This requires that the sub-contractor familiarises himself with all the contract documents; he cannot use the absence of an item from one contract document, when it exists in another, to argue for additional remuneration.

However, what is the position where the same item of work is differently described in the drawings and specification? If any work shown on the contract drawings is inconsistent with the description of that work contained within the contract specification or schedules of work then the contract drawings shall prevail for this purpose. The words 'shall prevail for this purpose' mean that the contract drawings will prescribe the quality and quantity of work included in the contract sum. It does not necessarily mean that the sub-contractor is obliged to perform the work shown on the contract drawings. This is a different issue, and where such an inconsistency exists the contractor shall issue directions in regard to the correction of that inconsistency (clause 2.4).

Therefore, it is possible that the sub-contractor will be required to perform the work shown in the specification but that the sub-contract sum will be adjusted on the basis of what is shown on the contract drawings.

With quantities (clauses 3.2 and 3.3)

Quantities can occur in two separate ways: either by the inclusion of a bill of quantities or by identification of the quantities required in the specification, schedule of works or other documents.

Where a bill of quantities is included in the sub-contract documents, the quality and quantity of the work included in the sub-contract sum is that which is set out in those bills of quantities. Where bills of quantities exist they should be included in the sub-contract documents. In the words of Donald Keating in *Building Contracts*:

'It is sometimes a difficult question of construction to determine whether quantities form part of the contract. The mere fact that quantities are submitted to the contractor for the purposes of tender does not make them form part ...'

Therefore, it is important to ensure that where bills of quantities exist they are incorporated into the sub-contract in order that they may prescribe the quality and quantities of work.

Where bills of quantities are to form part of the contract documents they are to be prepared in accordance with the sixth edition of the Standard Method of Measurement of Building Works, unless it is stated otherwise. If the work is not stated as being measured other than in accordance with SMM6 and there is a departure from the required method of measurement, the contractor is required to issue directions with regard to their correction. The correction may require a financial adjustment to be made to the sub-contract sum.

If no bills of quantities are used but quantities are nevertheless referred to in the numbered documents, the quality and quantity of work included in the sub-contract sum is taken to include that which is set out in those items which have been quantified. The effect of this is to put the responsibility for the quantification of the work upon the contractor, who in turn can look to the employer. If no such quantification is given in the specification, schedules and the like then the responsibility for quantifying the work is left with the sub-contractor. Where quantities are expressed in the specification or the like there is no requirement that the work shall be measured in accordance with SMM6. So, although bills of quantities must be measured in accordance with SMM6 unless expressly stated otherwise, quantities which are included in other documents can be measured in any way. Care should therefore be taken in pricing such items.

Interim payments

Clause 19 is the primary clause dealing with payment to the sub-contractor and sets out the timing of such payments and what amounts should be included in the payment.

With regard to the timing of interim payments, it is interesting to note that clause 4.2 of the IFC 84 states:

'Subject to any agreement between parties as to stage payments ... the supervising officer shall, at intervals of one month, unless a different interval is stated ... certify the amount of interim payments ...'

whereas NAM/SC in common with DOM/1 requires that interim payments shall be at monthly intervals and the amount of that payment shall subject to any agreement as to stage payments, include certain specified amounts.

The wording of IFC 84 is superior to that of NAM/SC and DOM/1 because these latter forms strictly speaking create the ludicrous situation of providing for interim payments on a monthly basis, but where stage payments have been agreed the interim payment is nil until the stage is achieved. This is a purely academic point and will not generally cause a problem in practice, but it is a good example of how words in one form of contract are better used than those in another.

Unfortunately, the same cannot be said for the remaining part of clause 4.2 in IFC 84 which reads 'certify the amount of interim payments to be made by the Employer to the Contractor within 14 days of the date of the certificate'. This part of the clause is badly drafted. The intention is that the supervising officer will certify amounts and that the employer will pay these amounts to the contractor within 14 days of the certificate.

In order to make sense, clause 19.2.1 which reads:

'the first interim payment shall be due no later than one month after the date of commencement of the sub-contract works on site or if so agreed of off-site works related thereto'

must be prefaced with 'subject to any agreement between the sub-contractor and contractor as to stage payments'. Generally, interim payments will be the norm as regards payment on account. This is because NAM/T does not provide specifically for any alternative arrangement. Although item 14 of section 1 or item 5 of section 11 may be used to this end, it seems unlikely that this will be the case in practice.

As this is so, and because once a sub-contractor is named the main contractor is obliged to contract, there seems little opportunity to agree that stage payments should be operative. What, then, is the incentive for the parties to agree to stage payments once the contract has been entered into? It may be advantageous to the main contractor but it seems most unlikely that it would be an advantage to the sub-contractor. In any event, even if an agreement was reached is there any consideration for such an agreement? Would the agreement be binding?

Timing of monthly interim payments

It can be seen from clause 19.2.1 that it is important to establish from which date the month runs. Is it the date for commencement which is included in the sub-contract documents, the date of commencement on-site notified by the contractor to the sub-contractor or the date upon which the sub-contractor actually commences on-site? It appears that the month should be taken from the date when the sub-contractor actually commences the sub-contract works on-site unless, of course, the contractor has agreed that the month shall run from the date of commencement off-site of the sub-contract works. Therefore, the dates given for commencement on or off-site have no effect. From a sub-contractor's point of view it is preferable that such an agreement takes place prior to contract, because afterwards there is little incentive for the contractor to agree.

One favours the view that time runs from the actual date of commencement on-site because it seems most appropriate and in keeping with the words used in the clause '. . . commencement of the sub-contract works on-site . . .'. However, one cannot be dogmatic about this because 'on-site' may be used purely to distinguish itself from 'off-site', and therefore it may be argued that the date notified for commencement on-site or, if agreed, off-site is the operative date regardless of whether the sub-contractor physically commences the work.

Whether the month runs from commencement on-site or off-site it is essential to record the date in order that the due date of the first and subsequent interim payments can be calculated. This will generally be a straightforward matter when commencement on site is used but where commencement off-site is the critical date the sub-contractor should inform the contractor accordingly, otherwise there could easily be a subsequent dispute.

The due date of each interim payment can therefore be established using the appropriate commencement date of the sub-contract works. These dates should not be confused with the valuation date nor the date when payment is due. In accordance with clause 19.3 the amount of each interim payment shall be ascertained at a date not more than seven days before the date when the interim payment is due. This date is known as the valuation date. Payment of the amount due shall under clause 19.2.3 be made not later than 17 days after the date when the payment becomes due. The following illustrates the time scale.

Commencement on site (sub-contract)	17 March 1987
Valuation date (earliest)	10 April 1987
Interim payment due	17 April 1987
Payment to be made (not later than)	4 May 1987

The valuation date will not necessarily correspond with the main contract valuation date. As a consequence, the main contractor may find himself contractually obliged to pay a sub-contractor before he has been paid himself. This is a situation towards which main contractors are not well disposed, but under the naming provision of IFC 84 it is very difficult (although not impossible) to secure a 'pay when paid' provision in the sub-contract.

Valuation of monthly interim payments

It is not essential that the sub-contractor gives notice of what he requires to be included in the interim payment, although it is necessary that certain documents are sent to the main contractor prior to final payment.

The sub-contractor is not contractually bound to give notice of what he requires to be included in the interim payment because clause 19.4 sets out exactly what should be included. No reference is made in the sub-contract as to who should carry out the valuation, but as the contractor is contractually obliged to pay such amounts as are referred to in clause 19.4 the onus clearly falls upon the main contractor. Nevertheless, the sub-contractor may feel that it is prudent to make such a request and to present a detailed valuation of the works executed and other relevant items. This may secure an enhanced valuation. The valuation is to include (clause 19.4.1):

(1)   total value of work properly executed.
(2)   total value of variations.
(3)   formula price adjustment – fluctuations (if applicable).
(4)   total value of materials and goods on-site.

The valuation may also include:

(5)   the value of materials and goods off-site.

A retention of 5% is held on all the above where the sub-contract work has not reached practical completion, but this is reduced to 2.5% for interim payments after practical completion.

In addition to the foregoing, the following amounts (clause 19.4.2) should also be included in the valuation and these amounts are to be included gross as they are not subject to retention.

(6) additional payment and costs arising from statutory obligations, fees and charges.
(7) ascertained loss and expense.
(8) contribution, levy and tax fluctuations (if clause 33 is operative).
(9) deductions in respect of errors in sub-contract work which are not required to be remedied.

The amount of the interim payment shall be the total of all the amounts referred to in (1) to (9) above less 'the amount calculated as due under clause 19.4 for the last previous interim payment ...' (clause 19.3.1) and where appropriate a 2.5% cash discount (clause 19.3.2).

The build-up of the valuation is fairly familiar but nevertheless there are still a number of points that require comment (it is not the intention to discuss these at length).

The ascertainment of the total value of work properly executed is generally considered to be straightforward. In order for this to be a straightforward exercise in legal terms, one must have a clear definition of how the value of work is to be calculated and a means of providing such a calculation. A priced bill of quantities, specification or schedule of rates, although essential, does not in itself achieve this end unless the total value is said to be ascertained by using this priced document. NAM/SC in common with most other forms of contract (save GC/Wks/1) does not prescribe this and therefore its use to this end can only be said to be convention. The contractor will be the arbiter of what work is properly executed, and if the sub-contractor disagrees his only recourse is to arbitration. The valuation of variations is fully discussed in the author's book *Variations in Construction Contracts*.

If the value of materials and goods is to be included in the valuation they must be reasonably and properly on site or adjacent the works. What these words mean is uncertain, particularly in the light of the other important aspects of clause 19.4.1.2. That is, the materials and goods must be

- for incorporation in the works
- adequately protected against weather and other casualties
- not brought to site or adjacent the site prematurely

Presumably 'reasonably and properly' refers to the condition of the goods and materials which have been delivered.

The contractor is obliged to include in the valuation the value of off-site materials and goods where the supervising officer has exercised his discretion to include such goods under clause 4.2.1(c) of the main contract conditions. All that clause 19.4.1.3 requires is that the contractor treats the sub-contractor in the same way as the contractor is treated by the employer.

The supervising officer has absolute discretion as to whether to include the value of materials and goods off-site. This discretion is regardless of the fact that it may have been agreed that the first interim payment is due one month from commencement off-site. The interim payment would become due, but if all the work executed to that date was off-site the supervising officer might still decline to include its value in the interim payment.

There is no appeal against the supervising officer's decision except to the client himself. As the supervising officer is under no obligation to include such amounts in the main contract valuation, the contractor and sub-contractor cannot rely upon such materials and goods being included in the valuation. If there were good grounds for such materials being included, it would be necessary for this to be established and agreed prior to contract. This would be a matter for the contractor to agree with both the sub-contractor and the employer. Otherwise, the contractor might find himself liable to the sub-contractor but without the same benefit being available from the employer.

The discretionary nature of the supervising officer's right to include off-site materials and goods is emphasised both in clause 4.2.1(c), IFC 84 and clause 19.4.1.3, NAM/SC where 'value' is used as compared with 'total value'. Clearly, the supervising officer can, in the exercise of his discretion, decide to include only part of the off-site materials and goods.

Retention

Clause 19.4 spells out those items upon which retention is in effect held and those upon which no retention is held. This is done by prescribing that prior to practical completion of the sub-contract works 95% of the value of the specified items should be included for interim payment, and that for interim payments thereafter it shall be 97.5%.

It should be noted that an effective reduction in retention takes place once an interim payment is made after practical completion of the sub-contract works. Whether this refers to the due date of the interim payment or the interim payment itself one cannot be sure, but it does seem on balance that if an interim payment is to be made after practical completion it should be at the 97.5% rate. This is in spite of the fact that either the valuation or the due date for the interim payment or both precede the practical completion date.

A 2.5% retention is held until the final payment is made in accordance with clause 19.8 and no provision, such as that made in JCT 80 and DOM/1, is made for the release of this retention upon the issue of the certificate of completion of making good defects.

There are no rules governing the creation of a trust for the monies retained and therefore the monies held by the main contractor would not necessarily be passed on to the sub-contractor in the event of the main contractor's insolvency. There would be no preferred status in respect of the retained amount and such sums would be mingled with the other sums available for general distribution to the creditors of the main contractor. This is notwithstanding the fact that the main contractor has the benefit of such a trust status for retention held on him by the employer under the main contract.

Previous payments

The calculation of the interim valuation is made gross, and each month a valuation for the whole of the works etc. is made in accordance with clause 19.4. From this amount the amount, if any, of the last interim payment is deducted. The amount of the last interim payment is, although it does not expressly say so, the amount due before the deduction of any cash discount (see example opposite).

Discount

If the contractor pays the sub-contractor the amount of the interim payment not later than 17 days after the date when the interim payment becomes due he is entitled under clause 19.3.2 to deduct 2.5% cash discount from the amount due and arising under clause 19.4.1. This indicates another good reason for honouring precisely the date on which the interim payment is due. The cash discount can be applied to all the amounts arising under the interim payments as referred to in clause 19.4.1. No distinction is made between the various elements of the valuation and therefore the cash discount may be applied equally to formula price adjustment as to work executed, but it cannot be applied to amounts arising under clause 19.4.2 and these amounts are to be paid gross.

The sub-contractor should be aware of the existence of the cash discount provisions although it is something that may be easily overlooked when tendering for work under NAM/SC provisions. This is so because although it is expressly referred to in the sub-contract there is only an oblique reference to cash discounts in NAM/T where reference is made with regard to daywork percentages.

Example: Calculation of interim payments

	Month One		Month Two	
Gross valuation as cl. 19.4.1	£100,000.00		£200,000.00	
Valuation to be paid at 97.5%		£97,500.00		£195,000.00
Gross valuation as cl. 19.4.2	£2,000.00		£4,000.00	
Valuation to be paid at 100%		£2,000.00		£4,000.00
		£99,500.00		£199,000.00
Less previous		—		£99,500.00
payments		£99,500.00		£99,500.00
Less 2.5% cash discount on £97,500 only if payment in accordance with cl. 19.2.3		£2,437.50		£2,437.50*
Amount to be paid		£97,062.50		£97,062.50

Gross valuation cl. 19.4.1		£200,000.00	
97.5%	=	£195,00.00	
Gross valuation cl. 19.4.2		£4,000.00	
		£199,000.00	
2.5% cash discount on £195,000		£4,875.00	
		£194,125.00 =	£97,062.50 +
			£97,062.50

*cash discount on £97,500, i.e. increase in valuation under clause 19.4.1 only.

On-site and off-site materials

The inclusion of materials on-site and off-site in interim payments has been discussed earlier in this chapter under the heading of interim payments. However, there are other clauses in the sub-contract and main contract which are concerned with materials, and these clauses are now dealt with separately.

Once materials and goods which are intended for the works are brought on to or adjacent to the site they should not be removed except for use in the works unless the contractor consents to their removal. A contractor's consent must be given in writing and any request by the sub-contractor to remove such materials shall be granted unless there is good reason for not doing so. Clause 19.5.1 is drafted to give this effect but it does seem that the provisions covering the giving of consent to remove such materials are superfluous. The reasoning is as follows:

(1)   if the materials are not intended for the works the sub-contractor has absolute discretion as to their removal. (Materials not intended for the works would cover those which are used for temporary works and materials not intended for either the main works or the temporary works. Clause 19.5.1 does not distinguish between the two, but presumably it is implied that the site cannot be used for the dumping of any materials that the sub-contractor so chooses.)

(2)   if materials are superfluous to requirements they are no longer required for the works.

(3)   the only materials not covered by (1) and (2) are those intended for the works. It is difficult to conceive a situation where such consent would be sought, and even more difficult to conceive a situation where such consent would be given.

(4)   if such an unlikely situation did arise a separate agreement could be sought.

Although it may be somewhat academic it should be noted that clause 19.5.1 is subject to clause 1.10 of the main contract conditions. The net effect of this is that before a contractor can agree to a sub-contractor removing materials intended for the works from the site, the contractor must first have secured the approval of the supervising officer.

Where a sub-contractor's materials have been included in a main contract interim payment under clause 4.2.1 of IFC 84 and the amount of that payment has been paid by the employer to the main contractor, the property in those goods is intended to pass to the employer. Clause 19.5.2 provides for this in an attempt to overcome the problem of an employer paying for such goods and then not finding himself the owner.

This problem was well illustrated by the case of *Dawber Williamson Roofing Ltd* v. *Humberside County Council* (1979) where it was decided that the sub-contractors had retained ownership of the materials notwithstanding a clause in the main contract to the effect that the ownership in materials which have been paid for by the employer was vested in the employer. The employer therefore lost out by paying for the materials, and yet did not have any legal right to their ownership because the terms of the main contract were not binding upon the sub-contractor. Clause 19.5.2 of NAM/SC purports to rectify this position and goes on to say that:

'... the sub-contractor shall not deny that such materials or goods are and have become the property of the employer'.

This means that once the employer has paid the contractor upon a certificate for the sub-contractor's materials, the sub-contractor cannot claim that the ownership of those materials is vested in him. This would give the employer some protection where the sub-contractor had a good title to pass on to the employer but not otherwise. Therefore, where a sub-contractor does give up ownership of materials as provided in clause 19.5.2 he is at risk until he is paid by the contractor.

As explained earlier, it is possible that the main contractor will pay the sub-contractor before he has himself been paid by the employer. Clause 19.5.3 deals with such a situation and provides that where a sub-contractor has received payment for his materials or goods then the property in those goods shall pass to the contractor.

From the foregoing it can be seen that it is the intention that once sub-contractor's materials or goods have been paid for the property in those materials or goods passes from the sub-contractor. However, this can only occur if the sub-contractor himself has ownership of the materials. The materials which have been supplied to the sub-contractor may be subject to a retention of title and therefore no title in those goods can be passed on. Therefore, where this occurs the person paying for such goods would still be at risk because he cannot secure title to those goods.

Clause 19.5.2 raises an important issue. Is it intended that such property should upon payment by the employer to the main contractor pass from the sub-contractor to the main contractor and on to the employer? Or is it intended to pass directly from the sub-contractor to the employer? Clause 19.5.3 expressly provides for the property to pass from the sub-contractor to the employer but clause 19.5.2 does not so provide. It is important to establish how the title in property passes because if it is through the main contractor the employer may have a remedy against the main contractor if such property did not pass. If, on the other hand, the title passed directly from the sub-contractor to the employer then it would seem that although the sub-contractor agrees he 'shall not deny that such materials or goods are and have become the property of the employer', this does not mean that he has passed on a good title. This is because, as already mentioned, the goods which have been sold to the sub-contractor may be the subject of a retention of title clause.

NAM/SC provides in clause 19.5.4 that the operation of clauses 19.5.1 to 19.5.3 shall be without prejudice to the property in any materials or goods passing to the employer in accordance with clause 1.11 of IFC 84. Clause 1.11 deals with off-site materials as clauses 19.5.2 and 19.5.3 are equally applicable to both materials on-site and materials off-site. It is important that the operation of these clauses does not prejudice clause 1.11 which in itself imposes further obligations upon the contractor.

Non-payment by main contractor

The contractor is responsible for paying to the sub-contractor the amount as calculated in accordance with clause 19. However, this amount may be the subject of a set-off (see pages 182–190) for a discussion with regard to set-off). If a named sub-contractor does not receive payment there is little point in his looking to the employer for such payment. The failure of the contractor to discharge a payment to the sub-contractor under the terms of the sub-contract is of no contractual concern to the employer, and in this respect NAM/SC and ESA/1 differ significantly from the NSC/4 and NSC/2 conditions of contract. The sub-contractor's only remedies are against the main contractor.

The contractual options open to the sub-contractor in the event of non-payment are:

- suspension of sub-contract works – clause 19.6
- determination by the sub-contractor of his own employment – clause 28.1.3
- pursue the contractor under his general rights in law

In any event, the contractor foregoes his 2.5% cash discount.

Both clauses 19.6 and 28.1.3 are without prejuduce to any other rights and remedies which the sub-contractor may possess and therefore the third option above is maintained and can be used in addition to the right to suspend or determine.

Clause 19.6 provides that where the contractor has failed to discharge his obligation to pay in accordance with the contract the sub-contractor may give notice of his intention to suspend the further execution of the sub-contract works. If after seven days from the date of notice the main contractor has still failed to pay then the sub-contractor may suspend his work until payment is discharged.

The sub-contractor should note that he must send a copy of his notice to the supervising officer. The supervising officer has no status in the matter and the notice is simply for information; nevertheless, the sub-contractor must comply with this requirement otherwise the validity of his notice to the contractor could be challenged. It is also necessary to serve a separate notice in respect of each default of payment.

Once the contractor has failed to pay the sub-contractor in accordance with the contract he foregoes his cash discount. Therefore, in order for the contractor to subsequently discharge his obligations he must pay the full amount exclusive of the cash discount. Where the sub-contractor has validly suspended his work and the main contractor pays the amount outstanding but less the cash discount, the sub-contractor would if he chose be entitled to continue his suspension. Suspension may continue so long as the payment to which the sub-contractor is entitled remains undischarged. Further, if the contractor fails to pay in due time but nevertheless deducts the cash discount to which he is no longer entitled the sub-contractor would be entitled to commence the process under clause 19.6.

Where a suspension of the sub-contract works arises as a consequence of the operation of clause 19.6, it is not the fault of the sub-contractor. Clause 19.6 states this specifically, but it does seem that this is obviously the intention and therefore such words are strictly unnecessary. Perhaps more important is whether such suspension would be considered to be the fault of the contractor, because if it would then the sub-contractor would be entitled to claim, in the absence of other contractual provisions, that 'time is at large' and that damage should be paid in respect of the breach. It is unclear in law whether such a suspension would be considered to be a fault of the contractor. This is because the *causa causans* is the non-payment by the contractor whereas the suspension is instigated by the sub-contractor himself.

It is therefore arguable whether the suspension was caused by the contractor. In order to clarify the position the sub-contract contains provisions which cover this situation:

(1)   Clause 12.7.14 – extension of time for delay caused by a suspension under clause 19.6.

(2) Clause 14.2.8 – loss and expense for disruption to the regular progress caused by a suspension under clause 19.6.

The extension of time is only available in respect of the sub-contract and therefore the responsibility for the delay, if any, to the main contract works is something which the main contractor has to accept. If a delay did occur to the main contract the employer would secure his building late, but his remedy is against the main contractor for liquidated and ascertained damages. With regard to any loss and expense paid to the sub-contractor, this is simply a matter for the contractor; the employer is not affected in any way.

Suspension of the sub-contract works under clause 19.6 may not always be appropriate. It may be that suspension is not necessary, not possible or that some more severe action is required. The sub-contractor may, for example, consider suing the contractor or seeking to determine his own employment. Both of these issues are dealt with separately under the chapter on determination and other remedies.

Final account and payment

Although it is not contractually necessary for the sub-contractor to submit details for the purposes of interim payments it is necessary for the purposes of the final account. Clause 19.7.1 provides.

> 'Either before or within a reasonable time after practical completion of the sub-contract works the sub-contractor shall send to the contractor all documents reasonably required for the purposes of the adjustment of the sub-contract sum.'

The sub-contract sum shall be adjusted by making additions and deductions in respect of variations, fluctuations, those matters referred to in clause 19.4.2 and also by deducting any provisional sums. Therefore, the documents reasonably required for the purpose of the adjustment of these items are to be submitted by the sub-contractor to the main contractor.

It is submitted that the sub-contractor should submit these regardless of any request to do so from the main contractor. There is no contractual requirement that the contractor must ask for such documents nor that the sub-contractor is only responsible for submitting those documents requested. The clauses may encounter a different interpretation on account of the words 'reasonably required' which, it has been suggested, mean the contractor reasonably requires. But the words, it is submitted, are not so used in this context.

If documents are considered necessary for the purposes of adjustment they must be submitted 'either before or within a reasonable time of practical completion of the *sub-contract works* ...' This is notwithstanding the fact that final payment is dependent upon the final certificate issued under the main contract conditions.

A final account of the sub-contract works is prepared and an adjusted sub-contract sum is arrived at. The amount of the final payment to the sub-contractor is the amount so calculated less the last interim payment, being the amount before deduction of any cash discount. The contractor is required by clause 19.8.2 to inform the sub-contractor, before final payment is due, what the amount of the final payment is to be. The final account must be prepared in time to meet the final payment date. Generally, there should be adequate time for this to be done because final payment is due not later than seven days after the issue of the final certificate under clause 4.4.2 of the main contract.

The date when final payment is due should not be confused with the date by which payment is to be made. Clause 19.8.2 states that:

'The final payment shall be made within 14 days of the date that it becomes due.'

If the contractor makes payment within this time he is entitled to deduct 2.5% cash discount in respect of the value of work referred to in clause 19.4.1.1. This means that the final account must separate the amounts which are subject to the cash discount.

The final payment is established by the contractor but it can be challenged by the sub-contractor if he is not satisfied. Proceedings must be commenced by the sub-contractor before or within 10 days after the notice is given by the contractor stating the amount due, or the final payment itself if that should for any reason precede the notice. Otherwise, the final payment becomes conclusive in respect of:

(1) quality of materials and standard of workmanship where required to be to the reasonable satisfaction of the supervising officer. This will only apply to specific items which are reserved for the opinion of the supervising officer.

(2) effect has been given to all the terms of the sub-contract that require adjustments to the sub-contract sum.

The final payment is not conclusive in respect of:

(1) whether the sub-contractor has fulfilled all his other contractual obligations.
(2) accidental inclusion or exclusion of any item or any arithmetical error in any computation.

These exceptions are extremely wide ranging and severely restrict the effect of the final payment. If, for instance, the sub-contractor has not complied with the specification and the supervising officer has not reserved the matter for his opinion then the sub-contractor can still be sued for the breach of contract. Furthermore, if items have been accidently included or excluded from the final account they can be corrected. This is very open ended, and although the clause refers only to 'accidental' and not 'wrongly' it is possible to argue subsequently that certain amounts included or excluded should be corrected. With regard to the correction of arithmetical errors, this must surely relate to those which are on the face of the final account, in spite of the words 'in any computation'. These words if not restricted could even extend to the build-up of a price, and if this was so then the conclusiveness of a final payment would amount to nothing.

Valuation of variations and provisional sum work

Valuation by agreement

In clause 16.1 NAM/SC provides that the amount by which the sub-contract sum is to be adjusted on account of a variation or expenditure of a provisional sum may be agreed between the contractor and sub-contractor. The clause requires that such agreement must be prior to the sub-contractor complying with the contractor's direction, and if agreement is not forthcoming the rules set out shall apply. This only sets out the procedure for valuing these items in the absence of agreement, but there is nothing to prevent the parties from agreeing as to the amount to be adjusted at any stage prior to compliance or afterwards. In other words, the wording does not prevent an agreement taking place after compliance of the direction but it does ensure that in the absence of an agreement there is a mechanism that can be used for valuation which becomes operative once the work is commenced.

It can be seen that there is a presumption in clause 16.1 that agreement will be reached, and failing that, the rules of valuation will be applied. This is in marked contrast to many standard forms of contract, including NSC/4 and DOM/1, where the presumption is that the rules of valuation will apply unless otherwise agreed.

Under NAM/SC the rules of valuation are a fall back position and it seems sensible that where possible the value of variations and the expenditure of provisional sums is best agreed between the contractor and sub-contractor. Naturally, as with any similar situations there are difficulties, and here the parties may resist an agreement in the belief that they are better served by the rules of valuation. Therefore, it follows that the price the sub-contractor would agree to would at least equal the amount that would be determined by the rules of valuation and would most probably exceed such an amount.

Variations to the sub-contract work and the expenditure of provisional sums under the sub-contract work are of concern to the employer in that he will have to account to the main contractor in respect of these matters. The employer is not, however, concerned with any agreement that takes place between the contractor and sub-contractor because the employer is not bound under the main contract to pay this amount. This is a domestic matter between the contractor and sub-contractor, and unlike the NSC/4 situation it is of little concern to the employer because he is not bound to pay any amount so agreed.

Because this is so the contractor will, before agreeing with the sub-contractor as to the value of the variation or expenditure of the provisional sum, seek an agreement with the employer. The sums involved will inevitably differ, and this is anticipated. If the contractor cannot secure an agreement with the employer then the contractor is unlikely to agree an amount with the sub-contractor unless the contractor is confident that when the rules of valuation are applied under the main contract he will secure sufficient recompense.

Where a variation or expenditure of a provisional sum cannot be valued by rates in the priced document and the sub-contractor and contractor agree what the amount of valuation shall be, it may put the employer in a difficult situation. The supervising officer may likewise have no rates to apply to the variation and therefore be left trying on behalf of the employer to resist the agreed amount (plus the contractor's on-costs).

Valuation by rules of valuation

The rules of valuation as prescribed in clause 16.2 are identical to those contained in IFC 84. The basic intention of the rules of valuation in NAM/SC is to achieve the same overall result as those in SFBC. For a full discussion on the rules of valuation under the JCT Standard Form of Building Contract the reader should refer to chapter 5 of the author's book entitled *Variations in Construction Contracts*.

The intention of producing the same valuation would have been assisted by prescribing the same legal framework. However, this is not the case. The words used in the various clauses differ and therefore the clauses lend themselves to the possibility of different interpretations. One does not wish to be pedantic and therefore the following observations and comments are restricted to the more fundamental issues.

The rules of valuation in clause 16.2.1 make reference to a priced document, and this is identified in NAM/T section I as one of the following documents: priced specification, priced schedules of work, contract bills (these are priced), contract sum analysis or schedule of rates. The relevant document depends upon which option has been chosen and identified in the invitation to tender. It should be noted that only one document can exist as the priced document on a particular project and that this priced document will form the basis for valuing any variations.

Clause 16.2.2 requires that omissions shall be valued in accordance with the relevant prices in the priced document. There is mention in clause 16.2.6 that account should be taken of preliminary items and this applies equally to omission and additions.

Clause 16.2.3 is similar to the combined effect of clauses 13.5.1.1 and 13.5.1.2 (JCT 80 with quantities) in that it provides:

> 'for work of similar character to that set out in the priced document the valuation shall be consistent with the relevant values therein making due allowance for any change in the conditions under which the work is carried out and/or any significant change in the quantity of the work so set out.'

Where the work which is the subject of a variation is similar in character to the work set out in the priced document its valuation shall be consistent with the relevant values in that document. The reference to the valuation being 'consistent with the relevant values therein' is necessary because the clause not only deals with identical items but also with those which are similar where work cannot be valued at the actual rates because of a change in conditions and/or quantity. The rates clearly form the basis for such valuation and due allowance is made for any change.

What constitutes a due allowance will be the subject of argument, but the intention can be summarised as follows. It is the difference between the item as priced in the priced document and the varied item if the varied item had been priced in the priced document. In other words, the value of the varied item is that at which it would have been priced at tender stage, and in order to achieve this retrospectively we refer to the relevant prices in the priced document.

Clause 16.2.4 provides for a fair valuation to be made where:

(1) the work is not of similar character.
(2) the valuation does not relate to the varied work.
(3) the valuation cannot reasonably account for the varied work or liabilities directly associated with the instruction.

A fair valuation takes one outside the priced document because a fair valuation is used in those circumstances where the priced document cannot be used to ascertain the relevant price. What then is the price to be ascertained as a fair valuation?

- is it the average price that one would be expected to pay at the time of carrying out the work?
- is it the actual price of carrying out the work?
- is it the average price that one would have expected to pay if the work had been priced at the time of tender.

In order that we may have a clearer idea of what is intended it is perhaps necessary to look also to clause 16.2.5 which provides:

'where the appropriate basis of a fair valuation is daywork, the valuation shall comprise ...'

Therefore, it can be seen that the actual price of carrying out the work may constitute a fair valuation, although this is only intended to be used where appropriate. Presumably it is appropriate when the variation cannot be measured and valued. However, when it can be measured and valued it seems that it would not be inappropriate to price the item at rates appertaining at the time of execution of the works. This is the logic of reading these clauses together, but whether that is what is really intended can only be a matter for conjecture.

The sub-contract contains no express requirements concerning the submission of daywork sheets, but it is in the sub-contractor's interest to provide the contractor with such information as soon as possible. The sub-contractor should in advance of carrying out the varied work attempt to agree with the contractor whether valuation by dayworks is appropriate. This will enable him to accurately record the relevant details. However, it is unlikely that a contractor will freely agree to dayworks at this stage, preferring to await the outcome of his own negotiations with the supervising officer under the main contract. Despite this, the sub-contractor is probably going to be best served by providing daywork sheets in respect of all variations where the work is to be valued at a fair valuation.

The contractor is not bound to accept that dayworks will be the appropriate basis for a fair valuation because if the work can be measured in some other way then it is appropriate that it should be. Nevertheless, once dayworks have been recorded the sub-contractor is likely to argue that even if you measure the work involved the answer must be the same because you would be obliged to apply rates to the measured work which are deduced from the daywork sheets. In other words, the daywork record provides the data base for labour and materials constants. This argument has been used for some time, with varying degrees of success but generally without much avail. However, on account of the changed wording the argument seems much more compelling under NAM/SC and IFC 84 than it does under the JCT standard forms.

Where a variation substantially changes the conditions under which other work is executed the work so affected shall be treated as though it were a variation. This means this work will also be valued under the rules of valuation contained in clause 16.2.

Clause 16.2.7 makes it clear that no allowance shall be provided in the valuation of the variation on account of disruption to the regular progress of the works or for any other loss and expense which will be reimbursed under other conditions of the sub-contract. Clause 14 and clause 14.2.6 are particularly relevant in this respect.

Valuation where no prices are set out in the priced document

The situation where no prices are set out in the priced document is envisaged and provided for in clause 16.2.9, but only when the priced document is a contract sum analysis or a schedule of rates. This is because where any other document is used it should be priced and will therefore contain relevant prices for the purposes of valuing variations. However, if because of some oversight a priced document did not exist the effect would be the same as implementing clause 16.2.9.

Clause 16.2.9 provides that a fair valuation of the variations shall be made when a contract sum analysis or schedule of rates is used and where no relevant rates or prices are set out therein.

The wording of clause 16.2.9 is interesting because it states:

'... and relevant rates and prices are not set out therein so that the whole or part of clause 16.2.2 to 16.2.8 cannot apply ...'

Therefore, if the rules of valuation cannot in whole or part apply, a fair valuation is to be made.

Clause 16.2.5, which provides for a fair valuation by daywork, is intended to be read in conjunction with clause 16.2.4 but it can be used separately to provide a valuation in the circumstances now under consideration. The only reason to prevent the use of clause 16.2.5 in this way is that the contract sum analysis or schedule of rates does not set out the rates which are to be applied to the prime costs of daywork.

An interesting argument therefore emerges. Should a fair valuation under clause 16.2.9 be made when a fair valuation can be made by dayworks, and if it should would the valuation be any different to that achieved by dayworks? It is submitted that a distinction does exist and should be made. A fair valuation would be used where work could be measured and would be valued at reasonable rates at the time of carrying out the work, whereas dayworks would be used when the work could not be measured and would then reflect the contractor's actual costs as compared with average costs.

Chapter 12

Payment – set-off and claims

Background

The major question that has been posed over the years in connection with set-off concerns what amounts can be set-off against certified sums. And where provision is made in the contract for the set-off of a certain sum, does this preclude the equitable right of set-off in respect of other claims?

The case law relevant to set-off is somewhat confusing and is made more so by the various opinions expressed by legal commentators. In *Dawnays Ltd* v. *F.G. Minter Ltd* (1971) it was held that the only sums that were deductable by the contractor from the certified amount were sums which are liquidated and ascertained and agreed to be due. The wording of the relevant clause was:

> 'The contractor shall notwithstanding anything in this sub-contract be entitled to deduct from or set-off against any money due from him to the sub-contractor (including any retention money) any sum or sums which the sub-contractor is liable to pay to the contractor under this sub-contract.'

The words 'any sum or sums' appear to be all embracing but the court took the view that these sums had to be liquidated and ascertained and sums for which the sub-contractor is *liable*. And furthermore, that upon true construction of the contract the clause by implication excluded the equitable right of set-off.

In 1973 the case of *Gilbert-Ash (Northern) Ltd* v. *Modern Engineering (Bristol) Ltd*, which was not on a standard form, was decided and, it has been suggested, effectively overruled the *Dawnays* decision.

Lord Dilhorne said:

'The existence of these provisions in the contract lends no support to the contention that the amount certified is of a special sacrosanct character which must be paid without deduction. If these deductions can be made, as they clearly can, from the amount certified, the amount which clause 30(1) says the contractor is entitled to be paid, why should it be inferred that the contract impliedly, for there is nothing express, excludes reliance by the employer on his common law and equitable rights to counterclaim and set-off for the amount certified? I see no ground for any such inference.'

It can be seen from this that both courts considered that the equitable right of set-off may be excluded by implication, the only real difference being that in *Dawnays* it was decided that upon true construction of the contract set-off was excluded by implication whereas in *Gilbert-Ash* it was concluded upon the true construction of that contract that the equitable right of set-off was not excluded. The reason for the suggestion that the *Dawnays* decision is overruled by *Gilbert-Ash* is surely based upon the view that if in the later case it was considered that the existence of very specific provision enabling amounts to be deducted from certified sums did not by implication exclude the equitable right of set-off, what can?

Therefore, it is argued that if an identical case to that of the *Dawnays* case was currently before the courts it would be decided differently. This may well be true, but it would be determined by what is the true construction of the contract. It is therefore possible that the equitable right of set-off can still be excluded by implication.

This was clearly shown in *Mottram Consultants Ltd* v. *Bernard Sunley & Sons Ltd* (1974) where the House of Lords held that only two permissible deductions were provided for and therefore nothing else could be deducted. On the face of it this case seems to go against the *Gilbert-Ash* decision, but it was decided once again upon the true construction of the contract. This form of contract had in it a set-off provision which had been deleted by the parties, and the House of Lords took the view that when construing a printed form regard could be given to those parts deleted. It concluded that the intention was to prevent other sums from being deducted from the amount certified.

Although the equitable right of set-off may be excluded by implication, in order to avoid any confusion it is desirable that the contract expressly provides for this if it is the intention to exclude such a right. Fortunately, NAM/SC does just this in clause 21.4 by providing that:

'The rights of the parties to the sub-contract in respect of set-off are fully set out in these sub-contract conditions and no other rights whatsoever shall be implied as terms of the sub-contract relating to set-off.'

Clause 21.4 therefore makes it clear that the only terms relating to set-off are set out in the contract in clause 21. That is not to say that the foregoing discussion is irrelevant, because it is necessary to understand the background to the court's attitude towards set-off and the impact, for instance, of deleting clause 21.4.

Set-off under NAM/SC

The contractor is given the right under clause 21.1 to deduct from any money due to the sub-contractor under the sub-contract:

(1) any amount agreed by the sub-contractor as due to the contractor.
(2) any amount awarded to the contractor in arbitration or litigation which arises out of or under the sub-contract.

In addition to this, clause 21.2 enables the contractor to deduct the amount of any loss and expense actually incurred by the contractor by reason of a breach on the part of the sub-contractor. However, before this additional right of set-off can be exercised the contractor is required to:

(1) quantify in detail and with reasonable accuracy the amount of the set-off.
(2) notify the sub-contractor in writing specifying his intention to set off the amount quantified and the grounds on which the set-off is claimed.

Clause 21.2.1 only requires that the contractor quantifies in detail and with reasonable accuracy the amount of set-off. There is no requirement that this information should be given to the sub-contractor. Although it is hoped that the contractor would provide such details, there is nothing which requires him to do so.

The notice that the contractor is required to give under clause 21.2.2 must be given no less than 20 days before the money from which the amount or part thereof is to be set-off becomes due and payable to the sub-contractor. The crucial issue here is establishing when the amount becomes due and payable. Unfortunately, this is not as clear as it should be because interim payments become due in accordance with the prescribed time scale yet the amount due need not be paid until some 17 days after the due date. Therefore, one could conclude that the amount need not be payable until the seventeenth day. This is unlikely in the context of clause 21.2.2 to be the intention, and payable is probably meant to be taken as synonymous with due. It would have been advantageous if the word payable had been omitted and the ambiguity thus avoided.

Clearly, it is advantageous to the contractor to construe the 20 days from the date payable as opposed to date due, thus enabling him to give a later notice and also to delay payment in order that the set-off notice period can be satisfied. It is submitted that the notice must be received by the sub-contractor not less than 20 days from the date due.

The contractor is not bound by any notice he may give under clause 21.2.2 and may amend it in preparing his pleadings for an arbitration upon the validity of the set-off. Likewise, any amount which the contractor sets-off under clause 21.2 is without prejudice to the rights of either party to vary the amount in any subsequent proceedings.

Sub-contractor's right to resist set-off

The sub-contract provides for set-off arising in differing circumstances, and the sub-contractor's right to resist set-off are variable. Under clause 21.1 the contractor can set-off any amount awarded in arbitration or litigation in favour of the contractor and which arises out of or under the sub-contract. If the sub-contractor wishes to avoid set-off where arbitration or litigation has been commenced he must win the argument before the award or decision is made.

Clause 21.1 also provides that the contractor is entitled to set-off any amount agreed by the sub-contractor. A sub-contractor can therefore resist any claim for set-off not made under clause 21.2 by not agreeing to the amount claimed by the main contractor. If the main contractor has approached the agreement with regard to set-off informally he would now be obliged to proceed under clause 21.2 if he wished to pursue the matter. As a sub-contractor is not likely to be well disposed to agreeing the amount of set-off the contractor is probably best advised to proceed, if not directly with the procedure under clause 21.2 then as soon as it becomes reasonably clear that an agreement will not be forthcoming.

Where a main contractor gives notice under clause 21.2.2 and the sub-contractor disagrees with the amount intended to be set-off the sub-contractor may object. The sub-contractor has 14 days from the receipt by him of the notice in which to send a written statement of counterclaim. This counterclaim must be sent by recorded delivery or registered post and must quantify with reasonable accuracy his claim, which must arise out of the sub-contract.

The strict requirement of clause 22.1.1 is that the sub-contractor when choosing to object must send his written statement within 14 days of receipt of the contractor's notice. This written statement does not need to be received by the contractor within the 14 days but the contractor's time period in which he can submit a defence to any counterclaim runs from the receipt of the sub-contractor's statement. Because the time periods run from the receipt of notices it is preferable that such notices are sent by registered post in order to avoid any dispute with regard to the date of receipt or whether such a notice has in fact been received.

At the same time as sending the counterclaim the sub-contractor is required to give notice of arbitration and to request action by the adjudicator. The sub-contractor is also obliged to inform the main contractor that he has requested the adjudicator to act. The sub-contractor is required to send to the adjudicator a copy of the contractor's statement of set-off and the sub-contractor's statement and counterclaim.

Clause 22.1.1 refers to a 'written statement of any counterclaim', and this suggests that the sub-contractor's only right is to counterclaim and not, as an alternative, to defend himself against the claimed amount of set-off. However, clause 22.1.1.2 envisages a statement, which is presumably a defence to the contractor's claim, and a counterclaim, if any. This is what must be intended, for a right to counterclaim but with no right simply to object to the amount claimed on the grounds that it is wrong would most certainly be absurd.

A good defence to a claim for set-off is that the loss and/or expense has not actually been incurred. The fact that loss and expense will be incurred is not sufficient reason to justify a set-off. Clause 21.2 refers only to 'loss and/or expense which has actually been incurred'. This matter was considered in *Redpath Dorman Long Ltd* v. *Tarmac Construction Ltd* (1982), albeit based upon the 'blue form' of sub-contract. The wording of the relevant clause in this sub-contract was identical to NAM/SC, and it was held that the terms of the clause did not allow any claim for loss or expense which the contractor might incur in future. This does not, however, prevent a subsequent set-off taking place once the loss has been incurred.

It is interesting to note that the sub-contractor's statement and counterclaim made under clause 22 is not binding and that he may amend them when preparing his pleadings for arbitration.

Where the sub-contractor decides to implement the procedure under clause 22.1.1 and sends a written statement and counterclaim to the contractor, the contractor may within 14 days from the reciept thereof send by recorded delivery or registered post to the adjudicator and sub-contractor a defence to any counterclaim made by the sub-contractor. The defence is to take the form of a written statement which sets out briefly his reasons as to why the counterclaim is not valid. It is interesting to note that the word defence is used in clause 22.2, whereas no such mention is made in clause 22.1.1.

Although the sub-contractor implements clause 22.1.1 because he objects to the amount of set-off, this does not prevent the contractor from actually setting-off the amount in question at the time that payment is due. The contractor should of course comply with the requirements of clause 21, otherwise he will be in breach of contract in making a set-off on account of his other rights in law being excluded by virtue of clause 21.4. The decision by the adjudicator will not generally be made before such set-off is due to be made by the contractor. All that has happened is that the sub-contractor reserves his right to have the set-off adjusted at a later stage. Figure 12.1 indicates the timescale related to the set-off procedure and it can be seen that it is possible, albeit unlikely, to secure the adjudicator's decision before payment is made by the main contractor.

Any set-off by the contractor under clause 21.2 and any counterclaim by the sub-contractor under clause 22.1.1 is without prejudice to the further exercise of their rights under these clauses if and when further sums become due to the contractor. This ensures that where the contractor's set-off is greater than the current payment due he can still deduct the balance of this set-off in subsequent interim payments, and also that the sub-contractor may continue to defend himself.

In the circumstances where the contractor notifies his intention to set-off in accordance with clause 21.2, the sub-contrator is not bound to implement the adjudication process as provided in clause 22 even though he may disagree with the set-off. The sub-contractor's failure to implement the adjudication process cannot be taken as an agreement to the amount set-off. The sub-contractor still has his rights under the contract, which include a reference to arbitration or to the courts. Obviously, it is generally in the interest of the sub-contractor to

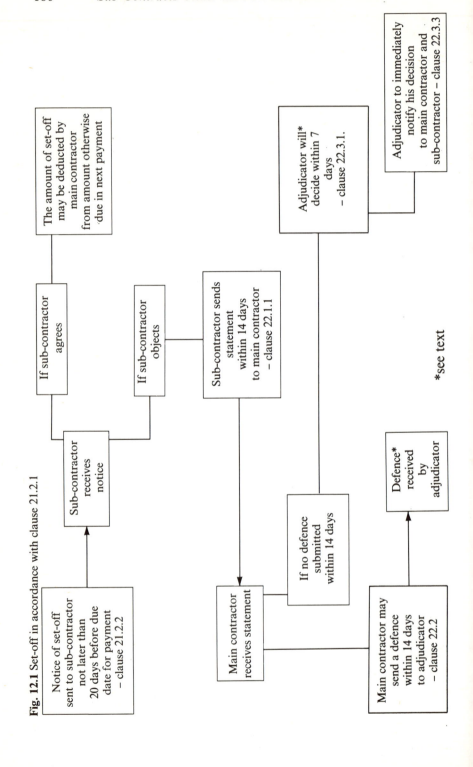

Fig. 12.1 Set-off in accordance with clause 21.2.1

implement the adjudication process; otherwise, the amount set-off may remain with the contractor for an inordinate length of time.

The adjudicator

The adjudicator is the person named in article 3 of section III of NAM/T. If the adjudicator is unable or unwilling to act then the person he appoints will act as the adjudicator. If no adjudicator is named then agreement would need to be reached as to who should adjudicate before clause 22 could operate.

Once a dispute has arisen, such agreement is going to be more difficult to achieve and is also going to be time consuming in circumstances when time is of the essence. A similar situation could also be reached if the appointed adjudicator decided not to act and then failed to appoint a successor. It is therefore very important that an adjudicator is appointed at the time of entering the sub-contract and that he undertakes either to carry out any such adjudication or to appoint someone who will. No person who has an interest in the sub-contract, main contract or in other contracts in which the sub-contractor or main contractor are involved can act as adjudicator unless it is agreed otherwise by the parties.

The sub-contractor can, subject to the provisions contained in the contract, request the adjudicator to decide those matters upon which the adjudicator has jurisdiction, namely those amounts notified by the contractor under clause 21.2.2.

It is desirable that the appointment of the adjudicator is made jointly by the sub-contractor and the main contractor and that the adjudicator is put under an obligation to make his decision in accordance with the time scale prescribed in clause 22.3.1; otherwise, the parties may be without remedy should the adjudicator fail to act or act unreasonably. However, the way the set-off procedure is arranged the appointment of the adjudicator is mainly for the benefit of the sub-contractor. No doubt that is why the sub-contractor is actually responsible for the adjudicator's fee. This does not mean that the sub-contractor will not recover from the contractor a part or the whole of the fee paid. The arbitrator has to settle responsibility with regard to the fee as well as the claim for set-off and may decide that it should be paid in full or part by the contractor.

When the main contractor submits a written statement under clause 22.2 the adjudicator has seven days from the receipt of that statement in which to make his decision. Where no such statement is made the adjudicator's decision is required within seven days from the expiry of the 14 day period referred to in clause 22.2. The decision of the adjudicator is to be made without any further statements, unless these are necessary to explain any ambiguity, and without any personal hearing.

The adjudicator is required to consider the whole amount set-off by the contractor under clause 21.2 and then, having considered the written statements provided under clause 22, make his decision which is to be fair and reasonable. The adjudicator's decision is, subject to clause 22.3.1, at his absolute discretion and is given without reasons. The decision is binding upon both parties until the matter is either settled by agreement or decided by an arbitrator or the court. The arbitrator appointed to settle the matter of the set-off may at the request of either party vary or cancel the adjudicator's decision at any time before making his final award.

The options available to the adjudicator when making his decision are set out in clause 22.3.1, and the amount of set-off may be dealt with as follows:

- retained by the contractor
- pending arbitration, held by a trustee-stakeholder
- paid to the sub-contractor
- a combination of these possibilities

For example, the contractor may have set-off a sum of £1000.00 and the adjudicator may decide that:

(1)   £500 shall be retained by the contractor.
(2)   £300 shall be paid by the contractor to the trustee-stakeholder.
(3)   £200 shall be paid by the contractor to the sub-contractor.

In this example the adjudicator has for the sake of clarity spelt out that £200 should be returned to the sub-contractor. This is not absolutely necessary as his decision in (1) and (2) implies that the balance should be returned. There is no possibility that the adjudicator may use the option in clause 22.3.1.3 other than to return amounts of set-off retained. Although it may appear otherwise, it does not enable the adjudicator to decide that the amount of any counterclaim in excess of the amount of set-off should be paid by the contractor to the sub-contractor.

Once the adjudicator has come to his decision, which should be within the prescribed time period, he is required to immediately notify it in writing to the contractor and sub-contractor. Here again it should be recognised that the adjudicator is required to perform under the sub-contract. He is not a party to this contract and therefore any remedy for default on his part must be sought under the contract between contractor and/or sub-contractor and the adjudicator, or in tort, or in both contract and tort.

The options available to the adjudicator under NAM/SC are the same as those contained in NSC/4, NSC/4a and DOM/1. But it should be noted that they are different from those contained in the Blue Form of Contract; although this provides certain options, only one could be adopted at any one time.

Execution of adjudicator's decision

Any amount of set-off which the adjudicator decides shall be paid to the sub-contractor shall be paid by the contractor immediately upon receipt of the adjudicator's decision. Where the adjudicator decides that an amount should be deposited with a trustee-stakeholder the amount shall be paid by the contractor. This amount is to be paid to the trustee-stakeholder 'thereupon' the adjudicator's decision.

Unfortunately, thereupon is an imprecise word which has a variety of meanings, but in the context of this clause it seems to mean 'soon' as opposed to 'immediately'. This interpretation is reinforced by the use of the word 'immediately' in clause 22.4.1. Such a distinction seems unnecessary and it is debatable whether it was intended.

Under either of the circumstances referred to, the amount paid by the contractor cannot in total exceed the amount due for payment under clause 19.3 to which the set-off relates. This is an unlikely situation, but it can occur where the contractor decides to pay further amounts to the sub-contractor even though the adjudication process has been commenced, or where the set-off exceeds the amount of the interim payment.

The trustee-stakeholder

Like the adjudicator, the trustee-stakeholder should be named in article 3 of section III of NAM/T, and for similar reasons it is important that this is done before entering into the contract.

NAM/SC sets out the position and role of the trustee-stakeholder, but his relationship with the parties to the sub-contract is not governed by the contract. Clause 22.5.1 provides that the trustee-stakeholder shall hold any sums received on account of the adjudicator's decision. The trustee-stakeholder holds the money in trust for the parties and holds it until such time as either the arbitrator or the parties themselves direct otherwise. The clause also provides that any sums so held shall be held on deposit and that interest shall be added. For the information of the sub-contract parties the clause also refers to the ability of the trustee-stakeholder to deduct his reasonable charges.

Where the trustee-stakeholder is a deposit-taking bank, clause 22.5.2 states that sums held can be held as an ordinary bank deposit and that these shall attract the usual rate of interest. This is done 'notwithstanding the trust imposed' and in effect attempts to change the responsibilities of the trustee-stakeholder which are imposed by statute. It is dubious whether this can be done.

Chapter 13

Determination

Generally

NAM/SC makes provision for the determination of the employment of the sub-contractor in three distinct ways:

- by the contractor (clause 27)
- by the sub-contractor himself (clause 28)
- automatically (clause 29)

Not all contracts incorporate such wide provisions and reference to GC/Wks/1 or the ICE Conditions of Contract will be sufficient proof of this. In these two forms of contract the contractor's only right to determine are those which exist in common law, there being no express provisions contained in the conditions.

Under NAM/SC the contractor's and sub-contractor's common law rights are preserved because the contract contains clauses stating that the determination provisions are without prejudice to these rights. The advantage of express conditions which provide for determination is twofold. Firstly, it prescribes the situations which enable determination to take place, whereas at common law it has to be established whether the breach entitles the party suffering as a consequence of that breach to determine the contract. An example of the problems of determining even where express conditions exist is illustrated in the case of *J. M. Hill and Sons Ltd* v. *London Borough of Camden* (1981). The employer, the London Borough of Camden, sought to determine the contract on account of the contractor cutting down his labour force on site. They submitted that this was a repudiatory breach of contract which entitled them to determine under clause 25 of their contract. It was held that they were not so entitled.

Secondly, the express provisions may include matters which would never be considered as sufficient to allow determination as a common law right. In essence, the common law position with regard to determination is that the party wishing to determine must show that the other party has either repudiated his obligations or has breached a condition that goes to the very root of the contract. In common law one talks of determination of the contract as compared with determination of the employment of the sub-contractor. If there is determination of the contract then the party entitled to determine will seek damages in respect of the breach and the loss flowing from the determination. The contract itself is at an end. If, however, there is determination of the employment of the sub-contractor then the contract still remains in existence. It is only the employment that is determined, and the rights and obligations of the parties in this event are as set out in the contract provisions.

The position is summarised in Vincent Powell-Smith and Michael Furmston's *A Building Contract Casebook* where they say:

> 'Where one party is entitled to determine the contract at common law, he will also be entitled to damages to compensate him for his loss. Where a party is terminating under an express provision of the contract, he will only be entitled to such further remedy as the contract gives him.'

This was deduced from the case of *Thomas Feather and Co. (Bradford) Ltd* v. *Keighley Corporation* (1953). However, it should be made clear that if a party determines under the express provision of the contract and the breach is also of a kind that would have entitled the party to determine even in the absence of these provisions then he may be entitled to common law damages. This case only decided that the breach which had occurred was not of the kind that would in the absence of express provisions to the contrary, have enabled determination to take place and therefore the only remedy available was that expressed in the contract. Further, the ability to determine for some reason which would not entitle a common law determination enables a party to determine without himself being pursued for damages on account of that determination.

Determination of employment of the sub-contractor

By the contractor

Clause 27 provides for determination on account of the specified defaults of the sub-contractor and also establishes the rights and obligations of the parties in the event that the sub-contractor's employment is determined.

The defaults which entitle the main contractor to determine the sub-contractor's employment are if the sub-contractor:

(1) without reasonable cause wholly suspends the carrying out of the sub-contract works before completion.

(2) without reasonable cause fails to proceed with the sub-contract works in accordance with clause 12.1 (this refers to progress in accordance with the agreed programme).

(3) after having received notice in writing, refuses or persistently neglects to remove defective work or improper materials or goods and this materially affects the work.

(4) after having received notice in writing, wrongfully fails to rectify defects and the like for which he is responsible under the sub-contract.

(5) fails to comply with the contract provisions in respect of sub-letting and/or fair wages.

The case of *Thomas Feather & Co. (Bradford) Ltd* v. *Keighley Corporation* (1953) established that failure to comply with the sub-letting provisions would not have entitled the corporation to determine in common law. Lord Goddard by way of *obiter dicta* felt that this was likely to be true in the case of the failure to comply with the fair wages provision and might be in respect of the other conditions. Therefore, the extent to which events contained in these provisions also give rise to the entitlement to determine under common law is extremely debatable. The contractor's right and remedies in common law are preserved by the existence of clause 27.4, but if he wishes to exercise such a right he must be absolutely certain that the breach goes to the root of the contract or that the sub-contractor can be said to have repudiated his obligations.

Even where a party seeks to determine under the express provisions of the contract he must be absolutely certain that he is entitled to do so. The consequences of a wrongful determination can be extremely costly.

The case of *J M Hill & Sons Ltd* v. *London Borough of Camden* has shown that reliance on such a clause is not straightforward, and that even in circumstances where the contractor reduces his workforce it does not entitle the employer to determine. Therefore, if determination under the express provision is sought one must ensure that:

(1)   a default of the kind specified in the contract has, in fact, occurred.
(2)   the contractor sends a notice by registered post or recorded delivery, specifying the default.
(3)   where the default persists for a further 10 days following receipt of the notice or the default is repeated, the contractor gives notice of determination by registered post or recorded delivery within 10 days of the continuance or repeat of the default.
(4)   he does not act unreasonably or vexatiously.

The words 'unreasonably or vexatiously' were considered by Lord Justice Lawton in the *Hill* v. *Camden* case and the learned judge said: 'I am far from clear as to what kind of conduct ... would make what he did unreasonable or vexatious.' In this case the London Borough of Camden argued that the only reason they had not paid the contractor was because he had wrongfully withdrawn men from site, and therefore for the contractor to determine on account of non-payment was indeed unreasonable. The judge, in spite of his earlier doubts, found that the plaintiff's action in determining his employment was not unreasonable nor vexatious. Had the contractor been at fault in withdrawing his men from site, it is possible and indeed likely that his subsequent actions would have been seen as unreasonable.

It is discretionary as to whether a contractor issues a notice of default and, subject to certain pre-conditions, discretionary as to whether he then determines the employment of the sub-contractor.

The contractor's right to exercise his discretion to determine commences 10 days after the receipt of the notice specifying the default, so long as the default is still continuing at this time, or immediately upon the repeat of such default. The contractor then has 10 days in which to exercise that right, and if he fails to do so he has waived his right under the express provision of the contract but not necessarily under common law.

Where the sub-contractor ceases his default following receipt of the notice from the contractor within the 10 day period the contractor is prevented from determining the sub-contractor's employment. However, if the sub-contractor repeats that default at any time, regardless of whether it has been previously repeated, the contractor can immediately give notice of determination. This means that once a valid notice of default has been issued the sub-contractor must be extremely careful not to repeat that default, because even where the contractor chooses to ignore a subsequent default he can upon a further repeat forthwith determine. The default repeated must be the same as the default specified in the notice in order to give the contractor the right to determine immediately. It is therefore a repeat of a specific default and not simply a further default.

A strange possibility occurs as a consequence of these provisions in that where a sub-contractor stops his default and then repeats it a determination may ensue more quickly than where the default just continues. Furthermore, if the default is continuous and still continuing more than 10 days after the notice of default then the contractor apparently has 10 days from the cessation of the default in which to determine. If the contractor fails to determine within this period then his rights to determine under the express provisions are extinguished. But of course, he still has his common law rights of determination and these may be used subsequently if it is appropriate. It has been suggested by Dennis Turner in *Building Sub-Contract Forms* that the 10 day period in which the contractor can determine commences from when the default ends, so long as the 10 day notice period has expired. This appears to be correct, but unfortunately the drafting of the clause is far from clear. The wording is a hangover from earlier contract forms and is perhaps surprising, bearing in mind that the Intermediate Form does not prescribe a time period in which the option to determine must be exercised.

By the contractor on account of bankruptcy or liquidation of sub-contractor

In the event of the sub-contractor becoming bankrupt or being put into certain specified situations which are likely to lead to bankruptcy or liquidation, the contractor may forthwith determine the employment of the contractor by giving written notices.

The events contained in clause 27.2 which give rise to this possibility are the sub-contractor:

(1) becoming bankrupt.
(2) the making of a composition or arrangement with his creditors.
(3) having a winding up order made against him.
(4) having a resolution for voluntary winding up passed (unless for the purposes of amalgamation or reconstruction).

(5)   having a provisional liquidator appointed.
(6)   having a receiver or manager of his business appointed.
(7)   having possession taken, by or on behalf of debenture holders who have benefit of a floating charge, of any property comprised in or subject to the floating charge.

Clause 27.2 provides that the contractor may determine if any of these events occur. The clause follows the DOM/1 and the 'blue' and 'green forms' of sub-contract in this respect, therefore overcoming many of the disadvantages of providing for automatic determination upon the occurrence of one of these events. NSC/4 and 4a provide for automatic determination in these circumstances, and the disadvantages of this provision are well described in John Parris's *Standard Form of Building Contract: JCT 80*. However, in consideration of these disadvantages it should be noted that NAM/SC does not make any reference to the assignment of any agreements between the sub-contractor and suppliers of materials.

Contractor's rights and obligations

The contractor's rights and obligations upon the determination of the sub-contractor's employment under clauses 27.1 and 27.2 are set out in clause 27.3.1. The position as stated in this clause is without prejudice to any arbitration or legal proceedings concerning the validity of the determination.

Once determination has occurred under the provisions of clause 27, the contractor is not bound to make any further payment to the sub-contractor until the sub-contract works are complete. The contractor may employ and pay others to complete the sub-contract works, and this includes the purchase of materials and goods. The contractor is entitled to use the sub-contractor's property for the purpose of completing the works.

Clause 27.3.2 seems to imply that the contractor may use anything that the sub-contractor has brought on to site which is intended for the sub-contract works. This right cannot extend to anything that may be hired in by the sub-contractor, because in the absence of any agreement to the contrary the hirer is perfectly entitled to the immediate return of his property. In practice this may cause a problem because the contractor may wish to continue the hire, and he will have to agree to pay for such hire. There is no certainty that such agreement will be forthcoming and provisions such as clauses 53(3) and 53(4) of the ICE Conditions of Contract would be desirable. Although the contractor would be entitled to charge such costs against the defaulting sub-contractor, there might be insufficient funds available for satisfying his claim.

When the contractor has finished with the sub-contractor's property he may ask for its removal. The sub-contractor is obliged upon the written direction of the contractor to remove within a reasonable time anything that belongs to him. In addition, the sub-contractor is required to remove anything on site that may have been hired by him. If the sub-contractor fails to do this the contractor may remove the property and sell it for the benefit of the sub-contractor. The contractor is entitled to deduct his costs from any sum realised and is not to be held responsible for any loss or damage to the property which he removes.

Clause 27.3.1, like clause 27.3.2, purports to give the contractor rights that he cannot possess. The clause refers to plant etc. which is hired by the sub-contractor, and upon failure to remove it when directed by the contractor the contractor apparently has the right to sell for the benefit of the sub-contractor. This he cannot do, because it is not the sub-contractor's property to sell and this could only be done with the hirer's agreement.

Once again, perhaps something can be learnt from looking at the ICE Conditions clauses 53(2), (6), (7), (8) which deal with this matter in an appropriate and lawful way. Furthermore, it is also uncertain that the clause which purports to relieve the contractor of the responsibility for loss and damage to the sub-contractor's property caused by its removal exonerates the contractor from any loss or damage howsoever caused. If the contractor's negligence causes the damage or loss then there is still a *prima facie* case for damages.

Any sums that are realised from the sale of the sub-contractor's property are to be held for the benefit of the sub-contractor. The only sums that can be deducted from this amount are the costs involved in selling the property. The contractor has no prior claim against the balance of such sums realised in respect of his other claims arising under the contract. This means that where the determination arises as a consequence of liquidation or the like the money realised from the sale of the sub-contractor's property is, save the expenses of the sale, for the benefit of satisfying all of the sub-contractor's creditors.

The sub-contractor is required under clause 27.3.3 to pay the contractor the amount which the contractor is obliged to recover under clause 3.3.6(b) of the Intermediate Form. This is a strange clause in that

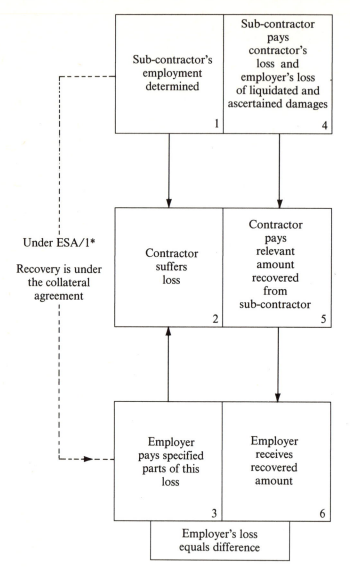

*This is illustrating a route of direct recovery available against the sub-contractor in respect of failure to design.

Figure 13.1

it attempts to make the sub-contractor responsible for certain additional costs caused by the sub-contractor's failure to perform, even though the party (the contractor) under his contract suffers no loss. The clause is therefore to protect the building employer against the losses he suffers as a consequence of having to pay the main contractor more and in respect of his inability to claim liquidated and ascertained damages from the contractor.

In the absence of such a clause in the sub-contract the sub-contractor would not be responsible for the losses of the employer as he is not a party to the contract. Clause 27.3.3 emphasises that the sub-contractor has notice of the main contract conditions and furthermore that the sub-contractor cannot contend that no loss has been suffered by the contractor on account of the operation of the main contract provisions by the employer. The overall effect of these provisions is illustrated by fig. 13.1.

Within a reasonable time of completion the sub-contract account should be drawn up by the contractor and verified. The final account shall set out:

(1)   the amount of direct loss suffered by the contractor on account of the determination together with all the other costs incurred in completing the sub-contract works (those already paid under 27.3.3 are excluded).
(2)   the amount (before deduction of the cash discount) paid to the contractor before determination.

The difference between the total of these amounts and the amount that would have otherwise been payable under the sub-contract is the amount which now has to be accounted for. Following precisely the wording of clause 27.3.4 the position, using example figures, could be taken as shown in Example 13.1. However, upon investigation the figures do not stand up and Example 13.2 illustrates this. The calculation shown in Example 13.2 needs to be made in order to arrive at the correct mathematical answer and to be in accordance with the contract wording.

What the contractor has to take account of is his loss of discount. The only way that this can be incorporated so that the strict words of the contract are adhered to is by including this loss of discount in the contractor's loss and expense. This is shown in Examples 13.3 and 13.4.

Example 13.1: Probable calculation in accordance with clause 27.3.4

	Example A £	Example B £
Loss and expense under cl. 27.3.4.1	4000.00	4000.00
Payments to others (excluding 27.3.3 payments already made) under cl. 27.3.4.1	36,000.00	36,000.00
Gross payments made to subcontractor before determination – cl. 27.3.4.2	160,000.00	160,000.00
Total amount	200,000.00	200,000.00
Amount that would have been paid to sub-contractor under original sub-contract cl. 27.3.4	180,000.00	204,000.00
Amount outstanding and to be paid by sub-contractor to main contractor	£20,000.00	
Gross amount outstanding and to be paid by main contractor to sub-contractor		4000.00
2.5% cash discount see cl. 27.3.4		100.00
Net Amount		£3900.00

Example 13.2: Calculation showing that Example 13.1 does not give an arithmetically correct answer

		Example A £		Example B £
Payment made to sub-contractor		156,000.00		156,000.00
Amount of contractor's loss and expense		4000.00		4000.00
Amount contractor pays to others to complete sub-contract		36,000.00		36,000.00
		196,000.00		196,000.00
Amount that would have been paid to sub-contractor under original contract	180,000.00		204,000.00	
less 2.5% cash discount	4500.00	175,500.00	5100.00	198,900.00
Additional cost/loss to contractor		£20,500.00		£2900.00

Example 13.3: Re-calculation of Example 13.2(A) to give arithmetically correct answer in accordance with clause 27.3.4

Example	£
Loss and expense under cl. 27.3.4.1	
loss as previously shown	4000.00
loss of discount as amount of sub-contract not executed at determination	
2.5% of (180,000–160,000)	500.00
Payments to others as before cl. 27.3.4.2	36,000.00
Gross payment made to sub-contractor before determination cl. 27.3.4.2	160,000.00
	200,500.00
Amount that would have been paid to sub-contractor under original contract cl. 27.3.4	180,000.00
Amount outstanding and to be paid by sub-contractor to main contractor	£20,500.00

Example 13.4: Re-calculation of Example 13.2(B) to give arithmetically correct answer in accordance with clause 27.3.4

Example	£
Loss and expense under clause 27.3.4.1	
loss as previously shown	4000.00
loss of discount on amount of sub-contract not executed at determination	
2.5% of (200,000–160,000)*	1025.64
Payments to others as before cl. 27.3.4.2	36,000.00
Gross payment made to sub-contractor before determination cl. 27.3.4.2	160,000.00
	201,025.64
Amount that would have been paid to sub-contractor under original contract cl. 27.3.4	204,000.00
	2974.36
Less 2.5% cash discount	74.36
Amount to be paid by main contractor to sub-contractor	£2900.00

*Note: add 1/39 as total amount subject to 2.5% deduction.

Determination of employment of the sub-contractor

By the sub-contractor

Clause 28 provides for determination on account of the specified defaults of the main contractor and also establishes the rights and obligations of the parties in the event that the sub-contractor determines his own employment.

The contractor's defaults which enable the sub-contractor to determine are if the contractor:

(1) without reasonable cause wholly suspends the works before completion.
(2) fails to proceed so that the sub-contract works are seriously affected.
(3) fails to make payment in accordance with the sub-contract.

The remedy of determination in the event of these defaults is only available if the default remedy under the other provisions of the contract would not adequately recompense the sub-contractor. For instance, a suspension of the works or the failure to proceed, thus causing the sub-contract works to be seriously delayed might be dealt with under clauses 12 and 14, the sub-contractor's remedies being an extension of time and loss and expense.

Therefore, because of the existence of these clauses and the wording of clause 28 the nature of the default must be very significant. Bearing in mind that in addition the sub-contractor must not act unreasonably or vexatiously, it is clear that in order to operate these provisions the default that occurs must be tantamount to repudiation. If this is so, little or no benefit is bestowed upon the sub-contractor by clauses 28.1.1 and 28.1.2 because under common law the sub-contractor already has these rights. The sub-contractor's other rights and remedies are protected because clause 28 is without prejudice to them.

With regard to the ability to determine on account of failure to pay, this is a useful provision. A sub-contractor would not necessarily possess a common law right to determine on account of non-payment and therefore this express provision is of importance. The contractor, knowing that the sub-contractor can determine his employment because of non-payment, will generally not be too reticent if the sub-contractor threatens him. But once again the sub-contractor has another option under the contract, and it may be considered that this should be used in the first instance: namely, suspension of the sub-contract works under clause 19.6.

The procedure for implementing the provision of clause 28.1 is exactly the same as for clause 27 (see pages 193–195.) except that the roles of the contractor and sub-contractor are reversed. However, if the sub-contractor has already decided to suspend the sub-contract works under the provision of clause 19.6 then the issue of the notice to determine cannot be given until 10 days after the commencement of the suspension.

It appears that in the first instance of non-payment by the main contractor the sub-contractor would be wise to implement clause 19.6 as opposed to clause 28. Once this has been done it is likely that the contractor will respond, because the remedy is given a reasonable chance to work in that the more drastic remedy of determination is automatically delayed.

Where a contractor persistently fails to pay the sub-contractor, the sub-contractor may well be entitled to such determination under clause 28 without first using the procedure contained in clause 19.6. Therefore, it can be seen that the sub-contractor's right to determine for non-payment, in practical terms, is verging on what in all probability are his common law rights.

Sub-contractor's rights and obligations

The sub-contractor's rights and obligations upon the determination of his own employment under clause 28.1 are set out in clause 28.3 and are without prejudice to his other rights and remedies in law. The position as stated in this clause is without prejudice to the accrued rights or remedies of the parties or to any liability arising from the indemnity given to the contractor by the sub-contractor in respect of injury to persons and property which may accrue before the sub-contractor shall have removed his property from site or upon removal from site.

Upon determination the sub-contractor shall remove from site all his temporary buildings, plant, tools, equipment, goods or materials. This he is required by clause 28.3.1 to do with all reasonable dispatch, but upon determination it is likely that the sub-contractor will immediately remove all of his property because commercially it is the prudent thing to do. The removing is required to be 'in such manner and with such precautions as will prevent injury, death or damage of the classes in respect of which he is liable to indemnify the contractor under clause 6'. This surely goes without saying, because the sub-contractor has obligations in law regardless of these provisions and would be unwise to act in any way that is improper or unreasonable.

Although the sub-contractor is required to remove all his things from site, this requirement is subject to clause 28.3.2.4. Presumably, the purpose of this is to try to ensure that any materials or goods brought on to site for the sub-contract works are not removed thus prejudicing the building employer. However, clause 28.3.2.4 does not exactly achieve this because it will only apply to materials which have been paid for by the contractor and where the property in those goods has transferred to the contractor. Admittedly, it can also apply to other materials which the sub-contractor is legally bound to accept upon such payment by the contractor but, it should be observed, there is nothing which specifically gives the contractor the right to purchase and therefore prevent removal. In fact the wording of clauses 28.3.1 and 28.3.2.4 does not really achieve their intention with respect to the restriction on the removal of goods because the latter clause is primarily dealing with account and not entirely compatible with clause 28.3.1.

There are other complications in respect of the operation of this clause caused by the possible retention of title by the suppliers. The problem can be summarised in the following senario:

(1)   materials or goods are delivered to site for which the sub-contractor is legally bound to accept may be subject to retention of title.
(2)   the main contractor pays the sub-contractor for such goods.
(3)   the materials or goods are purported then to be the property of the main contractor.
(4)   the sub-contractor fails to pay the supplier.
(5)   the result is that the property cannot pass to the main contractor because of the existence of a retention of title clause.

There is nothing to prevent an arrangement whereby the contractor pays the supplier direct for those goods, and where this is so a good title could pass to the main contractor. But without the appropriate provisions this course of action could be somewhat hazardous, and it is not intended by these provisions.

The contract provisions concerning these materials and goods are unduly confusing and ambiguous, mainly because of the right to purchase materials and the transfer of property being incorporated into a clause which is primarily one of account.

The sub-contractor's rights to payment upon determination are clearly set out in clause 28.3.2 but there is no mention of when this payment should be made. It is in the sub-contractor's interest to establish these amounts at the earliest possible moment and to provide an account to the contractor. Unfortunately for the sub-contractor, the contractor may decline or delay payment and the sub-contractor may be forced to take other proceedings.

Automatic determination

If the employment of the main contractor is determined under the specific provision of clause 7 of the main contract conditions then the employment of the sub-contractor shall thereupon automatically be determined (see page 196).

Determination under clause 7 can be:

- by the employer
- automatically
- by the contractor

The employer may determine the main contractor's employment if he shall make any one or more of the specified defaults in clause 7.1 (IFC 84) or is corrupt in accordance with clause 7.3 (IFC 84) or on account of one of the neutral events such as force majeure as referred to in clause 7.8.1 (IFC 84).

Where the contractor becomes bankrupt etc. his employment is 'forthwith automatically determined'. Although the main contract may be reinstated by agreement, the sub-contract is automatically determined upon determination of the main contract. Therefore, a separate agreement would have to be reached with the sub-contractor if he was to continue with the sub-contract works.

The contractor may determine his own employment if the employer shall make one or more of the defaults referred to in clause 7.5.1 (IFC 84), the employer becoming bankrupt etc. (clause 7.6, IFC 84,) or on account of one of the neutral events referred to in clause 7.8.1 (IFC 84). From this it can be seen that:

(1) the contractor's employment is not automatically determined where the employer becomes bankrupt etc., and therefore differs from where the contractor becomes bankrupt etc.
(2) the neutral events allow either the employer or contractor to determine.

Where the sub-contract is automatically determined as a consequence of any of the above, the provisions of clause 28.3 shall apply. However, there is an important exception in that if the determination is on account of a neutral event referred to in clause 7.8.1 (IFC 84) then the sub-contractor is not entitled to recover under 28.3.2.6 any direct loss and expense as a consequence of the determination.

Automatic determination under clause 29 of the sub-contract only takes place if the main contract is determined under clauses 7.1, 7.2, 7.3, 7.5, 7.6 or 7.8 (IFC 84). The main contract clauses are without prejudice to the other rights and remedies the parties may possess. Determination under their common law rights is not in accordance with these clauses, and therefore there is no automatic determination of the sub-contract if they are exercised by the parties to the main contract.

Chapter 14

Fluctuations

Generally

Under NAM/SC, fluctuations may be adjusted using either clause 33 or clause 34. Clause 33 provides for fluctuations in contributions, levies and taxes whereas clause 34 provides for formula price adjustment . The choice of fluctuation provisions is made in item 16 of section I of NAM/ T. It is intended that one identifies which of these alternatives is to apply, but one should only use formula price adjustment where contract bills are used. It is possible, however, to adopt formula price adjustment without using bills of quantities but clearly this is not intended, nor is it desirable without making the necessary provisions in order that it can operate. Furthermore, one can choose not to adopt any fluctuation provision and delete all references to clause 33 and 34 in NAM/T. Again, this is not intended but a firm price contract can be achieved without too much difficulty.

Bearing in mind the likely duration of the majority of contracts let under the Intermediate Form of Building Contract, it does seem that the use of formula price adjustment will be the exception rather than the rule. Therefore, the likely choice is clause 33 which provides for adjustment to contributions, levies and taxes. However, considering the anticipated size of the contracts let under the form and the difficulties with implementing this clause economically, one would be tempted to delete both options provided.

Contributions, levies and taxes

Where clause 33 is operative it provides that the sub-contract sum shall be adjusted on account of changes to contributions, levies and taxes.

The contributions, levies and taxes referred to in clause 33 shall be taken to include:

'... all impositions, payable by a person in his capacity as an employer howsoever they are described and whoever the recipient which are imposed under by virtue of an Act of Parliament and which affect the cost to an employer of having persons in his employment.'

and in respect of:

'... types and rates of duty if any and tax (other than value added tax ...) by whomsoever payable which at the Date of Tender are payable on the import, purchase sale, appropriation, processing or use of the materials, goods, electricity ...'

Therefore, in order for adjustment to take place the contribution etc. must be imposed or by virtue of an Act of Parliament; but not, it seems, if it is payable under the Industrial Training Act 1964 which is specifically excluded in clause 33.1.2, or in respect of value added tax which is excluded by clause 33.5.2.

The sub-contract sum is deemed to be based upon the rate of contributions etc. payable at the Date of Tender which should be stated in item 16 of section I of NAM/T, or if not there then in item 4 of section II of NAM/T. This means that the sub-contractor is not required to take account of any proposed rate changes even if they are known at the date of tender. It is extremely important that the date of tender is inserted in NAM/T, otherwise there is going to be particular difficulty with any contributions, levies and taxes which have become payable around the time when tenders were submitted. Furthermore, the sub-contractor is not required to take account in his tender of any proposed new contributions if they are not payable at the date of tender, but he will be able to recover the cost of these contributions if they become payable by him after the date of tender.

The adjustment made to the contract in respect of contributions, levies and taxes is the net difference between what the sub-contractor pays and what he would have paid if no changes had taken place. The interpretation of net difference seems straightforward, but it has been implemented in a number of ways. It is submitted that, when dealing with contributions related to labour, the actual wage sheets should be used for the basis of calculation and the two sets of contributions applied to these figures. The difference together with any percentage addition under clause 33.7 is the amount by which the sub-contract sum is adjusted. Where the contributions relate to materials and goods the two sets of contributions should be applied to the invoice price. Again, the difference together with any percentage addition under clause 33.7 is the amount by which the sub-contract sum is adjusted.

The adjustment is not made in respect of all labour and materials used. It is restricted in a number of ways. Although most labour employed by the sub-contractor in connection with the sub-contract works falls within the scope of clause 33, it does exclude the following:

(1) those working on or adjacent to the site in connection with the sub-contract works who are not *workpeople*. Workpeople are defined in clause 33.6.3 as those whose wages and other emoluments (including holiday credits) are governed by the National Joint Council for the Building Industry or some other wage-fixing body for trades associated with the building industry. This definition therefore excludes 'lump labour' as their wages are not determined in this way. Confirmation of this position can be found in the case of *Murphy* v. *London Borough of Southwark* (1982). A definition of wage-fixing body is to be found in clause 33.6.4.

(2) workpeople engaged upon the production of materials and goods for use in or in connection with the sub-contract works not on site or adjacent to the site but who are not *directly* employed.

(3) those employed on or adjacent to the site and who are not defined as workpeople and have not been on site for at least two days in the week.

(4) head office staff.

(5) labour on daywork.

It should be noted that the calculation for adjustment in respect of those employed on or adjacent to site and who are not defined as workpeople but have been on site for two days or more is different in that the difference does not relate to their own wages but to those that are payable in respect of a craftsman.

Clauses 33.1.5 and 33.1.6 also provide that amounts which are received by the sub-contractor under or by virtue of an Act of Parliament in his capacity as an employer and which affect the cost to him of having persons in his employment are also adjustable.

Where the sub-contractor pays employers' national insurance contributions at the contracted-out rate it shall be ignored, and the sub-contractor will for the purposes of fluctuations be paid or required to pay the net amount which relates to the national insurance contributions at the non-contracted-out rates.

Fluctuations in respect of sub-let work

Clause 33.3 regulates the position with regard to fluctuation recovery in respect of sub-let work. The sub-contractor cannot require adjustment under clause 33.1 in respect of sub-let work because the contributions etc. are not paid by him in his capacity as an employer. However, the sub-contractor can require adjustment under clause 33.3.2 if he has incorporated the similar fluctuation provisions in the sub-sub-contract. The sub-contractor, where sub-letting under clause 24.2, is required under clause 33.3.1 to incorporate such clauses to the like effect of clause 33, and therefore should be able to require such an adjustment. If the sub-contractor fails to do so then neither himself nor the sub-sub-contractor can recover any adjustment for fluctuations. However, if the main contractor was unable to recover from the sub-contractor under clause 33.3.2 because of the sub-contractor's failure then the main contractor will have a claim for breach of contract.

Notice a condition precedent to payment

Clause 33.4.1 requires written notice of any event that will lead to an adjustment of the contributions, levies and taxes; and this notice, which must be given in a reasonable time, is a condition precedent to payment.

Certainly in practice such notice is seldom forthcoming, and therefore subsequent claims can be resisted. However, the sub-contractor could not rely on his own default to avoid paying any decreases in contributions.

In addition to this notice the sub-contractor may be required by the contractor to provide evidence and computations in order that the fluctuations may be ascertained. In respect of fluctuations arising from sub-sub-contracts and employees other than workpeople the sub-contractor must also certify the validity of the evidence supplied.

Amount of fluctuations payable

Clause 33.4.3 states that the contractor and sub-contractor may agree what shall be deemed for all the purposes of the sub-contract the net amounts of fluctuations. Ordinarily, the contractor and sub-contractor could agree what they liked in respect of the sub-contract as they are the parties to the contract and as such at liberty to vary the contract. Therefore, one may be inclined to the view that this clause is unnecessary. Interestingly, any such agreement made as to fluctuations will affect the employer by virtue of clause 4.10 (IFC 84). This leaves the employer open to possible abuse, even though the contractor and sub-contractor may be restricted only to determine quantum and not liability following the arguments expressed in *John Laing Construction Ltd* v. *County and District Properties Ltd* (1982).

It is fairly clear that the ability to determine the amount of fluctuations may well be very significant. Does it then follow that the agreement must be reasonable? There is nothing to indicate that this should be the case, nor much to indicate that the employer has much of an argument save the possibility of fraud.

The amounts which become payable or allowable to the sub-contractor in respect of fluctuations under clause 33 shall be added to or deducted from the sub-contract sum or the amount which is otherwise payable on determination under clause 28. This amount will not in any way alter the amount of profit that the sub-contractor has included in the sub-contract sum.

When the amounts are ascertained as fluctuations they should be included in full in interim payments and any outstanding amounts should be included in the final payment payable under clause 19.8.1, subject to clauses 33.4.5 and 33.4.7. This means that the sub-contractor must have provided evidence and computations as reasonably required by the contractor. It also means the sub-contractor is not entitled to fluctuations which arise from events which occur after the date so notified by the contractor under clause 13. This latter point is governed by clause 33.4.7 but is subject to clause 33.4.8.

Clause 33.4.8 provides that clause 12 is left unamended and that the contractor has given his decision in respect of every written notice for an extension of time given under clause 12.

Although the clause restricts recovery flowing from events occuring after the date when the sub-contractor should have completed, it does not restrict the recovery of all fluctuations. Fluctuations can still be recovered for work executed whilst the sub-contractor is in default, but only at the level appertaining at the date when he went into default.

The wording used in clause 33.4.7 attempts to prevent the sub-contractor securing increases that have only occurred as a consequence of his own default. It does seem, however, that this clause only applies to amounts which are included in interim payments and not to any amount which is included in the final payment. This clause has, it seems, been handed down from earlier forms. The wording in clause 4.7 of IFC 84 is preferable, referring as it does to both 'interim' and 'final' payments.

Formula adjustment

It is not intended to consider these provisions in detail here because:

(1)   they will only be used in a limited number of instances.
(2)   much detailed discussion as to the applications of formula price adjustment exists, albeit in respect of other contract forms; but the operation is virtually identical.

However, certain points with regard to formula price adjustment have already been considered in respect of the completion of NAM/T at pages 46–48.

A further point concerns clause 34.5 which enables the contractor and sub-contractor to agree to alter the methods and procedures for ascertaining the amount of formula fluctuations, but unlike clause 33 it is necessary that such an ascertainment must be reasonably expected to be the same or approximately the same as it would have been had the agreement not taken place.

Chapter 15

Settlement of disputes

NAM/SC makes provison for all disputes or differences concerning the sub-contract to be referred to arbitration. However, it is also possible to refer questions of law to the High Court but, this will depend on whether or not article 5.3 has been identified to apply in item 6 of section I of NAM/T. If article 5.3 has been identified to apply in NAM/T then it would not be necessary to show that it had in fact been adopted in the main contract form.

Nevertheless, it is debatable whether the clause stands up in law, because although the Arbitration Act 1979 provides for questions of law to be referred to the High Court where the parties consent it seems that such consent must relate to a question of law arising in the course of a reference. The adoption of Article 5.3 does not fulfil this requirement. However, where Article 5.3 is intended to apply and one party only wishes to refer the point to the High Court the arbitrator may be inclined to give his consent. This procedure would, it seems, satisfy the requirements of the Arbitration Act 1979, but as one can see, the commencement of an arbitration would be necessary and one cannot go directly to the Court.

Where a dispute or difference under the contract does arise it shall be referred to arbitration and the final decision of the arbitrator. The arbitrator is to be agreed by the parties, and they are given 14 days from the request to appoint in which to agree. Failing agreement, which is fairly common, either party may request a person to be appointed by the President or Vice-President of the Royal Institution of Chartered Surveyors.

The arbitrator is given wide ranging powers under the contract, in fact wider powers than possessed by the courts in respect of such matters as opening up and reviewing a supervising officers decision, as can be seen from *Northern Regional Health Authority* v. *Derek Crouch Construction*

Co. Ltd (1983). Nevertheless, the arbitrator is restricted in a number of ways and cannot arbitrate upon matters which are:

(1)   stated by the final payment as conclusive, unless proceedings have already commenced or are commenced within 10 days of the notice of final payment.
(2)   amounts agreed by the contractor and sub-contractor pursuant to clauses 33.4.3 and 34.5. These are references to the agreement as to what the fluctuations shall be deemed to be for the purposes of the sub-contract.

Where article 5.4 has been identified to apply in item 6 of section 1 of NAM/T, the joinder provisions will be operative. This is a sensible provision enabling disputes under the sub-contract which are substantially the same as those already referred under the main contract to be referred to the same arbitrator. Where this occurs the arbitrator is purported to be given powers to make such directions and awards in the same way as if the procedure of the High Court as to joining one or more defendants or joining co-defendants or third parties were available to the parties and to him. This sets out an intention which can only partly be enforced in that an arbitration agreement is only binding on those who have agreed to be bound by it, and therefore the arbitrator's power and the parties themselves are restricted as a consequence.

Table of Cases

Table of standard form contract clauses

Index